MAKERS OF ELECTRICITY

BROTHER POTAMIAN
JAMES JOSEPH WALSH

Édition et mise en page : Culturea (Hérault, 34)
Retrouvez notre catalogue sur http://culturea.fr/
Contact : infos@culturea.fr
Imprimé en Allemagne par Books on Demand Gmbh
Design typographique et layout : Derek Murphy et Reedsy
ISBN : 9791041985227
Date de parution : février 2024
Ce livre :
— a été composé avec une police open source libre de droits dessinée sur Open Fondry https://open-foundry.com/
— est édité en libre access sous licence CC0 (Creative Commons Zero)
afin de conserver l'oeuvre au plus près des caractéristiques du domaine public.
— offre la possibilité à son lecteur de partager, dupliquer ou distribuer son contenu, sans restriction ni réserve.
— permet de replanter un arbre. SHS Éditions soutient Plantons Pour l'Avenir, fonds de dotation pour le reboisement des forêts françaises
400 projets de reboisement de forêts soutenus depuis 2014 à travers la France https://www.plantonspourlavenir.fr/

CHAPTER I. PEREGRINUS AND COLUMBUS.

The ancients laid down the laws of literary form in prose as well as in verse, and bequeathed to posterity works which still serve as models of excellence. Their poets and historians continue to be read for the sake of the narrative and beauty of the style; their philosophers for breadth and depth of thought; and their orators for judicious analysis and impassioned eloquence.

In the exact sciences, too, the ancients were conspicuous leaders by reason of the number and magnitude of the discoveries which they made. You have only to think of Euclid and his "Elements," of Apollonius and his Conics, of Eratosthenes and his determination of the earth's circumference, of Archimedes and his mensuration of the sphere, and of the inscription on Plato's Academy, *Let none ignorant of geometry enter my door,* to realize the fondness of the Greek mind for abstract truth and its suppleness and ingenuity in mathematical investigation.

But the sciences of observation did not advance with equal pace; nor was this to be expected, as time is an essential element in experimentation and in the collection of data, both of which are necessary for the framing of theories in explanation of natural phenomena.

The slowness of advance is well seen in the development of the twin subjects of electricity and magnetism. As to the lodestone, with which we are concerned at present, the attractive property was the only one known to ancient philosophy for a period of six hundred years, from the time of Thales to the age of the Cæsars, when Lucretius wrote on the nature of things in Latin verse.

Lucretius records the scant magnetic knowledge of his predecessors and then proceeds to unfold a theory of his own to account for the phenomena of the wonder-working stone. Book VI. of "De Natura Rerum" contains his speculations anent the magnet, together with certain observations which show that the poet was not only a thinker, but somewhat of an experimenter as well. Thus he recognizes magnetic *repulsion* when he says: "It happens, too, at times that the substance of the iron recedes from the stone as if accustomed to start back from it, and by turns to follow it."

This recognition of the repelling property of the lodestone is immediately followed by the description of an experiment which is frequently referred to in works on magnetic philosophy. It reads: "Thus have I seen raspings of iron, lying in brazen vessels, thrown into agitation and start up when the magnet was moved beneath"; or metrically,

And oft in brazen vessels may we mark Ringlets of Samothrace, or fragments fine Struck from the valid iron bounding high When close below, the magnet points its powers.

This experiment, seen and recorded by Lucretius, is of special interest to the student of magnetic history because of the use which is made of iron filings and also because it has led certain writers to credit the poet with a knowledge of what is known to-day by the various names of magnetic figures, magnetic curves, magnetic spectrum. We do not, however, share this view, because we see no adequate resemblance between the positions assumed by the bristling particles of iron in the one case, as described by the Roman poet, and the continuous symmetrical curves of our laboratories in the other. If Lucretius noticed such curves in his brazen vessels, he does not say so; nor does the meagre description of magnetic phenomena given in Book VI. warrant us in assuming that he did.

The use of iron filings to map out the entire field of force that surrounds a magnet was unknown to classical antiquity; it was not known to Peregrinus or Roger Bacon in the thirteenth century or even to Gilbert in the sixteenth. The credit for reviving the use of filings and employing them to show the direction of the resultant force at any point in the neighborhood of a magnet, belongs to Cabeo, an Italian Jesuit, who described and illustrated it in his "Philosophia Magnetica," published at Ferrara in the year 1629. On page 316 of that celebrated work will be found a figure, the first of the kind, showing the position taken by the filings when plentifully sifted over a lodestone: thick tufts at the polar ends with curved lines in the other parts of the field.

The Samothracian rings mentioned in the passage quoted above were light, hollow rings of iron which, for the amusement of the crowd, the jugglers of the times held suspended one from the other by the power of a lodestone.

Writing of the lodestone, Lucretius says:

Its viewless, potent virtues men surprise, Its strange effects, they view with wond'ring eyes, When without aid of hinges, links or springs A pendent chain we hold of steely rings. Dropt from the stone—the stone the binding source— Ring cleaves to ring and owns magnetic force; Those held above, the ones below maintain; Circle 'neath circle downward draws in vain Whilst free in air disports the oscillating chain.

Though the Roman poet was acquainted with two of the leading properties of the lodestone, viz., attraction and repulsion, there is nothing in the lines quoted above or in any other lines of his great didactic poem to indicate that he was aware of the remarkable difference which there is between one end of a lodestone and the other. The polarity of the magnet, as we term it, was unknown to him and remained unknown for a period of 1200 years.

During that long period nothing of importance was added to the magnetic lore of the world. True, a few fables were dug out of the tomes of ancient writers which gained credence and popularity, partly by reason of the fondness of the human mind for the marvelous, and partly also by reason of the reputation of the authors who stood sponsors for them.

Pliny (23-79 A. D.) devotes several pages of his "Natural History" to the nature and geographical distribution of various kinds of lodestones, one of which was said to repel iron just as the normal lodestone attracts it. Needless to say that the mineral kingdom does not hold such a stone, although Pliny calls it *theamedes* and says that it was found in Ethiopia.

Pliny is responsible for another myth which found favor with subsequent writers for a long time, when he says that a certain architect intended to place a mass of magnetite in the vault of an Alexandrian temple for the purpose of holding an iron statue of Queen Arsinoe suspended in mid-air. Of like fabulous character is the oft-repeated story about Mahomet, that an iron sarcophagus containing his remains was suspended by means of the lodestone between the roof of the temple at Mecca and the ground.

As a matter of fact, Mahomet died at Medina and was buried there in the ordinary manner, so that the story as currently told of the suspension of his coffin in the "Holy City" of Mecca, contains a twofold error, one of place and the other of position. By a recent (1908) imperial irade of the Sultan of Turkey, the tomb is lit up by electric light in a manner that is considered worthy of the "Prophet of Islam."

Four centuries after Pliny, Claudian, the last of the Latin poets as he is styled, wrote an idyl of fifty-seven lines on the magnet, which contains nothing but poetic generalities. St. Ambrose (340-397) and Palladius (368-430), writing on the Brahmans of India, tell how certain magnetic mountains were said to draw iron nails from passing ships and how wooden pegs were substituted for nails in vessels going to Taprobane, the modern Ceylon. St. Augustine (354-430) records in his "De Civitate Dei" the wonder which he felt in seeing scraps of iron contained in a silver dish follow every movement of a lodestone held underneath.

With time, the legendary literature of the magnet became abundant and in some respects amusing. Thus we read of the "flesh" magnet endowed with the extraordinary power of adhering to the skin and even of drawing the heart out of a man; the "gold" magnet which would attract particles of the precious metal from an admixture of sand; the "white" magnet used as a philter; magnetic unguents of various kinds, one of which, when smeared over a bald head, would make the hair grow; magnetic plasters for the relief of headache; magnetic applications to ease toothaches and dispel melancholy; magnetic nostrums to cure the dropsy, to quell disputes and even reconcile husband and wife. No less fictitious was the pernicious effect on the lodestone attributed in the early days of the mariner's compass to onions and garlic; and yet, so deeply rooted was the belief in this figment that sailors, while steering by the compass, were forbidden the use of these vegetables lest by their breath they might intoxicate the "index of the pole" and turn it away from its true pointing. More reasonable than this prohibition was the maritime legislation of certain northern countries for the protection of the lodestone on shipboard. According to this penal code, a sailor found guilty of tampering with the lodestone used for stroking the needles, was to have the guilty hand held to a mast of the ship by a dagger thrust through it until, by tearing the flesh away, he wrenched himself free.

It was only at the time of the Crusades that people in Europe began to recognize the *directive* property of the magnet, in virtue of which a freely suspended compass-needle takes up a definite position relatively to the north-and-south line, property which is serviceable to the traveler on land and supremely useful to the navigator on sea.

It is commonly said that the compass was introduced into Europe by the returning Crusaders, who heard of it from their Mussulman foes. These, in turn, derived their knowledge from the Chinese, who are credited with its use on sea as far back as the third century of our era.

Among the earliest references to the sailing compass is that of the trouvère Guyot de Provins, who wrote, about the year 1208, a satirical poem of three thousand lines, in which the following passage occurs:

The mariners employ an art which cannot deceive. An ugly stone and brown, To which iron joins itself willingly They have; after applying a needle to it, They lay the latter on a straw And put it simply in the water Where the straw makes it float. Then the point turns direct. To the star with such certainty That no man will ever doubt it, Nor will it ever go wrong. When the sea is dark and hazy, That one sees neither star nor moon, Then they put a light by the needle And have no fear of losing their way. The point turns towards the star; And the mariners are taught To follow the right way. It is an art which cannot fail.

The author was a caustic and fearless critic, who lashed with equal freedom the clergy and laity, nobles and princes, and even the reigning pontiff himself, all of whom should be for their subjects, according to the satirist, what the pole-star is for mariners—a beacon to guide them over the stormy sea of life.

Guyot traveled extensively in his early years, but later in life retired from a world which he despised, and ended his days in the peaceful seclusion of the Benedictine Abbey of Cluny.

An interesting reference, of a similar nature to that of the minstrel Guyot, is found in the Spanish code of laws known as Las Siete Partidas of Alfonso el Sabio, begun in 1250 and completed in 1257. It says:

"And even as mariners guide themselves in the dark night by the needle, which is their connecting medium between the lodestone and the star, and thus shows them where they go alike in bad seasons as in good; so those who are to give counsel to the king ought always to guide themselves by justice, which is the connecting medium between God and the world, at all times to give their guerdon to the good and their punishment to the Wicked, to each according to his deserts."

It will be necessary to give a few more extracts from writers of the first half of the thirteenth century in order to show how little was known about the magnet and how crude were the early appliances used in navigation when Peregrinus appeared on the scene.

Cardinal Jacques de Vitry, who lived in the East for some years, wrote his "History of the Orient" between the years 1215 and 1220, in which he says:

"An iron needle after touching the lodestone, turns towards the north star, so that such a needle is necessary for those who navigate the seas."

This passage of the celebrated Cardinal seems to indicate that even then the compass was widely known and commonly used in navigation.

Neckam (1157-1217), the Augustinian Abbot of Cirencester, wrote in his "Utensilibus":

"Among the stores of a ship, there must be a needle *mounted on a dart* which will oscillate and turn until the point looks to the north; the sailors will thus know how to direct their course when the pole-star is concealed through the troubled state of the atmosphere."

This passage is of historical value, as it contains what is probably the earliest known reference to a mounted or pivoted compass. Prior to the introduction of this mode of suspension, the needle was floated on a straw, in a reed, on a piece of cork or a strip of wood, all of which modes of flotation, when taken in conjunction with the unsteadiness of the vessel in troubled waters, must have made observation difficult and unsatisfactory.

Brunetto Latini (1230-1294) makes a passing reference to the new magnetic knowledge in his "Livres dou Tresor," which he wrote in 1260, during his exile in Paris.

"The sailors navigate the seas," he says, "guided by the two stars called tramontanes; and each of the two parts of the lodestone directs the end of the needle that has touched it to the particular star to which that part of the stone itself turns."

Though a statesman, orator and philosopher of ability, the preceptor of Dante in Florence and guest of Friar Bacon in Oxford, Brunetto has not got the philosophy of the needle quite right in this passage; for the part that has been touched by the north end of a lodestone will acquire south polarity and will not, therefore, turn towards the same "tramontane" as the end of the stone by which it was touched.

Dante himself admitted the occult influence on the compass-needle that emanates from the pole-star when he wrote:

"Out of the heart of one of the new lights There came a voice that, needle to the star, Made me appear in turning thitherward. *Paradise*, XII., 28-30.

The next writer on the compass is Raymond Lully (1236-1315), who was noted for his versatility, voluminous writings and extensive travels as well as for the zeal which he displayed in converting the African Moors. Lully writes in his "De Contemplatione": "As the needle after touching the lodestone, turns to the north, so the mariners' needle directs them over the sea."

This brings us to the last of our ante-Peregrinian writers who make definite allusions to the use of the compass for navigation purposes, viz., Roger Bacon, one of the glories of the thirteenth century as he would be of the twentieth. It was at the request of his patron, Pope Clement IV., that Bacon wrote his "Opus Majus," a work in which he treats of all the sciences and in which he advocates the experimental method as the right one for the study of natural phenomena and the only one that will serve to extend the boundaries of human knowledge. In a section on the magnet, a clear distinction is drawn between the physical properties of the two ends of a lodestone; for "iron which has been touched by a lodestone," he says, "follows the end by which it has been touched and turns away from the other." Besides being a recognition of magnetic polarity, this is equivalent to saying that unlike poles attract while like poles repel each other. Bacon further remarks, by way of corroboration, that if a strip of iron be floated in a basin, the end that was touched by the lodestone will follow the stone, while the other end will flee from it as a lamb from the wolf. There is, however, an earlier recognition known of the polarity of the lodestone; for Abbot Neckam, fifty years before, called attention to the dual nature of the physical action of the lodestone, attracting in one part (say) by sympathy and repelling at the other by antipathy. It was the common belief in Bacon's time and for centuries after, that the compass-needle was directed by the pole-star, often called the sailor's star; but Bacon himself did not think so, preferring to believe with Peregrinus, that it was controlled not by any one star or by any one constellation, but by the entire celestial sphere. Other contemporaries of his sought the cause of the directive property not in the heavens at all, but in the earth itself, attributing it to hypothetical mines of iron which, naturally enough, they located in regions situated near the pole. Peregrinus records this opinion, which he criticises and rejects, saying in Chapter X. that persons who hold such a doctrine "are ignorant of the fact that in many different parts of the globe the lodestone is found; from which it would follow that the needle should turn in different directions, according to the locality, which is contrary to experience." A little further on he gives his own view, saying: "It is evident from the foregoing chapters that we must conclude that not only from the north pole (of the world), but also from the south pole rather than from the veins of mines, virtue flows into the poles of the lodestone."

Observations had to accumulate and much experimentation had to be done before it was finally established that the cause of the directive property of the magnet is not to be sought in the remote star depths at all, but in the earth itself, the whole terrestrial globe acting as a colossal magnet, partly in virtue of magnetic ore lying near the surface and partly also in virtue of electrical currents, due to solar heat, circulating in the crust of the earth.

Of the early years of Pierre le Pélérin (Petrus Peregrinus), nothing is known save that he was born of wealthy parents in Maricourt, a village of Picardy in Northern France. From his academic title of Magister, we infer that he received the best instruction available at the time, probably in the University of Paris, which was then in the height of its fame. His reputation for mathematical learning and mechanical skill crossed the Channel and reached Friar Bacon in the University of Oxford. In his "Opus Tertium," the Franciscan Friar records the esteem in which he held his Picard friend, saying: "I know of only one person who deserves praise for his work in experimental philosophy, because he does not care for the discourses of men or their wordy warfare, but quietly and diligently pursues the works of wisdom. Therefore it is that what others grope after blindly, as bats in the evening twilight, this man contemplates in all their brilliancy because he is master of experiment."

Continuing the appraisal of his Gallic friend's achievements, he says: "He knows all natural sciences, whether pertaining to medicine and alchemy or to matters celestial and terrestrial. He has worked diligently in the smelting of ores and also in the working of minerals; he is thoroughly acquainted with all sorts of arms and implements used in military service and in hunting, besides which he is skilled in agriculture and also in the measurement of lands. It is impossible to write a useful or correct treatise on experimental philosophy without mentioning this man's name. Moreover, he pursues knowledge for its own sake; for if he wished to obtain royal favor, he could easily find sovereigns to honor and enrich him."

This is at once a beautiful tribute to the work and character of Peregrinus and an emphatic recognition of the paramount importance of laboratory methods for the advancement of learning. It is evident from such testimony, coming as it does from an eminent member of the brotherhood of science, that the world had not to wait for the advent of Chancellor Bacon or for the publication of his *Novum Organum* in 1620, to learn how to undertake and carry out a scientific research to a reliable issue. Call the method what you will, inductive, deductive or both, the method advocated by the Franciscan friar of the thirteenth century was the one followed at all times from Archimedes to Peregrinus and from Peregrinus to Gilbert, none of whom knew anything of Lord Bacon's pompous phrases and lofty commendation of the inductive method of inquiry for the advancement of physical knowledge. Be it said in passing, that Bacon, eminent as he undoubtedly was in the realm of the higher philosophy, was, nevertheless, neither a mathematician nor a man of science; he never put to a practical test the rules which he laid down with such certitude and expectancy for the guidance of physical inquiry. Moreover, there is not a single discovery in science made during the three centuries that have elapsed since the promulgation of the Baconian doctrine that can be ascribed to it; it has been steadily ignored by men renowned in the world for their scientific achievements and has been absolutely barren of results.

Peregrinus, on the other hand, does not stop to enumerate opinions, he does not even quote Aristotle; but he experiments, observes, reasons and draws conclusions which he puts to the further test of experiment before finally accepting them. Then and then only does he rise from the order of the physicist to that of the philosopher, from correlating facts and phenomena to the discovery of the laws which govern them and the causes that produce them. Furthermore, he was in no hurry to let the world know that he was grinding lodestones one day and pivoting compass-needles the next; what he cared for supremely was to discover facts, new phenomena, new methods. Peregrinus was not an essayist, nor was he a man of mere book-learning. He was a clear-headed thinker, a close and resourceful worker, a man who preferred facts to phrases and observation to speculation.

At one period of his life, Master Peter applied his ingenuity to the solution of a problem in practical optics, involving the construction of a burning-mirror of large dimensions somewhat after the manner of Archimedes; but though he spent three years on the enterprise and a correspondingly large sum of money, we are not told by Friar Bacon, who mentions the fact, what measure of success was achieved. Bacon, however, avails himself of the occasion to insinuate a possible cause of failure, for he says that nothing is difficult of accomplishment to his friend *unless it be for want of means*.

Centuries later, the French naturalist Buffon took up the same optical problem, with a view to showing that the feat attributed to Archimedes during the siege of Syracuse by the Romans was not impossible of accomplishment. For this purpose, he used 168 small mirrors in the construction of a large concave reflector, with which he ignited wood at a distance of 150 feet and succeeded in melting lead at a distance of 140 feet. As this was done in the winter time in Paris, it was concluded that it would have been quite possible to set a Roman trireme on fire from a safe distance by the concentrated energy of a Sicilian sun.

If Peregrinus was alert in mind, he appears to have been very active in body. Prompted, no doubt, by the higher motives of Christian faith and perhaps a little, too, by his fondness for travel and adventure, he took the cross in early life and joined one of the crusading expeditions of the time. That he went to the land of the paynim, we have no direct evidence; but we infer the fact from the title of Peregrinus or Pilgrim, by which he is known, his full name being Pierre le Pélérin de Maricourt, or, in the Latinized form, Petrus Peregrinus de Maricourt.

In 1269, we find him engaged in a military expedition undertaken by Charles Duke of Anjou, for the purpose of bringing back to his allegiance as King of the Two Sicilies the revolted city of Lucera in Southern Italy. He served in what might be called the engineering corps of the army, and was engaged in fortifying the camp and constructing engines of defense and attack. Unlike his companions in arms, Peregrinus does not allow himself to be wholly absorbed with military duties, nor does he waste his leisure hours in frivolous amusements; his mind is on higher things; he is engrossed with a problem in practical mechanics which required him to devise a piece of mechanism that would keep an armillary sphere in motion for a time.

In outlining the necessary mechanism, as he conceived it, he was gradually led to consider the general and more fascinating problem of perpetual motion itself, with the result that he waxed somewhat enthusiastic when he thought that he saw the possibility of constructing an ever-turning wheel in which the motive power would be magnetic attraction, the attraction of a lodestone for a number of iron teeth arranged at equal distances on the periphery of a wheel. The device looked well on paper, beyond which stage it was not carried, perhaps for want of leisure, or more probably for want of the necessary material and tools. Had Peregrinus been able to test his theoretical views on the *magnetic motor* by actual experiment, the delusive character of perpetual motion would have been recognized at an early epoch in the world's history, and much time and money spared for more profitable investment.

This very wheel, which was designed in the trenches before Lucera in 1269, was probably the cause of the withering rebuke which Justin Huntly McCarthy administers in his "History of the French Revolution," Vol. I., p. 256, where he says: "In the long record of rascaldom from *Peregrinus* to Bamfylde Moore Carew, no single rascal stands forward with such magnificent effrontery, such majestic impudence, such astonishing success as Cagliostro." To say the least, this is a very serious slip of the pen on the part of the Irish historian of the French Revolution, in which a scientific pioneer of the first rank and a patriot of exalted type is mistaken for a charlatan of the deepest dye.

Although Peregrinus puts the burden of constructing his wheel on others, he does not appear to have considered it a vain conceit; for, in the beginning of the last chapter of the "Epistola" he says: "In this chapter, I will make known to you the construction of a wheel which, in a remarkable manner, moves continuously." He is writing from Southern Italy to his friend Siger (Syger, Sygerus), at home in Picardy; and that this friend may the better comprehend the mechanism of the wheel, he proceeds to describe in a systematic manner the various properties of the lodestone, all of which he had investigated and many of which he had discovered. The "Epistola" of Peregrinus is, therefore, the first treatise on the magnet ever written; it stands as the first great landmark in magnetic philosophy.

The work is divided into two parts—the first contains ten chapters and the latter three. "At your request," he says to his friend, "I will make known to you in an unpolished narrative the undoubted though hidden virtue of the lodestone, concerning which philosophers, up to the present time, give us no information. Out of affection for you, I will write in simple style about things entirely unknown to the ordinary individual."

After this declaration as to the original character of his work Peregrinus proceeds: "You must know that whoever wishes to experiment should be acquainted with the nature of things; he must also be skilled in manipulation, in order that by means of this stone, he may produce those marvelous results."

The titles of the chapters will give an idea of the comprehensive character of the magnetic work accomplished by the author and, at the same time, will serve to show how much was known about the lodestone in the thirteenth century.

PART I.

Chap. I. Purpose of this work.
II. Qualifications of the experimenter.
III. Characteristics of a good lodestone.
IV. How to distinguish the poles of a lodestone.
V. How to tell which pole is north and which south.
VI. How one lodestone attracts another.
VII. How iron touched by a lodestone turns towards the poles of the world.
VIII. How a lodestone attracts iron.
IX. Why the north pole of one lodestone attracts the south pole of another, and *vice versa*.
X. An inquiry into the natural virtue of the lodestone.

PART II.

Chap. I. Construction of an instrument for measuring the azimuth of the sun, the moon or any star when in the horizon.
II. Construction of a better instrument for the same purpose.
III. The art of making a wheel of perpetual motion.

An attentive reading of the thirteen chapters of this treatise of 3,500 words will show that:

(1) Peregrinus assigns a definite position to what he calls the *poles* of a lodestone and gives practical directions for determining which is north and which south.

(2) He establishes the two fundamental laws of magnetism, that like poles repel and unlike poles attract each other.

(3) He demonstrates by experiment that every fragment of a lodestone is a complete magnet, and shows how the fragments should be put together in order to reproduce the polarity of the unbroken stone.

(4) He shows how a pole of a lodestone may neutralize a weaker one of the same name and even reverse its polarity.

(5) He pivots a magnetized needle and surrounds it with a circle divided into 360 degrees.

This brief summary shows the great advance made by the author on what was known about the lodestone before his time. Most of the salient facts in magnetism are clearly described and some of their applications pointed out. So thorough and complete was this apprehension and explanation of magnetic phenomena that nothing of importance was added to it for the next three hundred years.

In the compass which Peregrinus devised for use in navigation, a light magnetic needle was thrust through a slender vertical axis made of wood, which axis also carried a pointer of brass or silver at right angles to the needle. According to the belief of the time, the magnetic needle gave the north and south points of the horizon, while the brass pointer determined the east and west points. This compass, double pivoted be it noticed, was provided with a graduated circle and a movable arm, having a pair of upright pins at its extremities, which movable arm enabled the navigator to determine the magnetic bearing of the sun, moon or any star at the time of rising or setting. "By means of this instrument," the author says in Chap. II., "you can direct your course towards cities and islands and any other place wherever you may wish to go, by land or by sea, provided you know the latitude and longitude of the place which you want to reach."

The invention of the compass has been attributed to one Flavio Gioja, a seafaring man of Amalfi, a flourishing maritime town in Southern Italy. If we admit that Gioja was a real and not a fictitious person, we cannot, however, admit the claim which is made by his countrymen, when they say that he gave to the mariner the use of the compass in the year 1302; for we have seen that Peregrinus distinctly states that his compass, described in 1269, could be relied upon for guidance by the traveler on land as well as by the voyager on sea.

To Gioja may belong the merit of having simplified and improved the compass. It is likely that he suspended the needle on one pivot instead of the two used by Peregrinus, and that he added the compass-card with its thirty-two divisions, attaching it to the needle itself, thereby adding materially to the practical character of the compass as a nautical instrument.

On the other hand, a claim has been made for Peregrinus which cannot be admitted. It was put forward by his itinerant countryman Thévenot, in the seventeenth century, to the effect that the author of the "Epistola" was acquainted with magnetic declination, in virtue of which a freely suspended magnet does not point north and south, but cuts the geographical meridian at a definite angle.

Writing in 1681, Thévenot says in his "Recueil de Voyages" that: "It was a matter of general belief down to the present day, that the declination of the magnetic needle was first observed sometime in the beginning of the last (16th) century. I have found, however, that there was a declination of five degrees in the year 1269, having found it recorded in a manuscript with the title "Epistola Petri Adsigerii," etc.

The title of the manuscript seen by Thévenot is not, however, as he gives it above, but "Epistola Petri ad Sygerium," etc., which is quite a different reading.

There are twenty-eight manuscript copies of the "Epistola" known to exist; and only one of them, that of the University of Leyden, contains the passage alluded to by Thévenot. This manuscript was the object of careful study and critical examination by Wenckebach (1865) and other competent scholars, who pronounced it a spurious addition made some time in the early part of the 16th century.

In the time of Peregrinus, it is probable that the declination did not exceed three degrees in Paris or on the shores of the Mediterranean, a quantity so small that it would have been difficult of detection; and, if detected, would have been attributed either to errors in the construction of the instrument used or to inaccuracy on the part of the observer. This is what happened to Columbus when, on his return to Spain, having reported the many and definite observations on the variation of the compass which he had made on his outward voyage, he was told by the learned ones of the day that *he* was in error and not the needle, because the latter was everywhere true to the pole.

This oft-stated and widely-believed fidelity of the needle to the pole is not, however, founded on fact; it is the exception, the rare exception, not the rule, despite the couplet of the poet:

Th' obedient steel with living instinct moves And veers for ever to the pole it loves;

or this other,

So turns the faithful needle to the pole, Though mountains rise between and oceans roll.

That the magnet does not turn to the pole of the world is common knowledge to-day, when the High School tyro will tell you that in New York it points 9° *west* of north, while in San Francisco it points 15° *east* of north. If he happens to be well up, he may refer to the position of the agonic line on the globe along which the needle stands true to the pole, while all places to the east of that line in our hemisphere have westerly declination and those to the west have easterly declination. Indeed, magnetic charts show places where the needle points east and west instead of north and south, and others where the north-seeking end points directly south. Such varying and conflicting behavior of the compass-needle serves to show the irregular manner in which the earth's magnetism is distributed and also the intensity of distributing forces which exist at certain places.

It is one of the gems in the crown of Columbus, that he observed, measured and recorded this strange behavior of the magnetic needle in his narrative of the voyage. True, he did not notice it until he was far out on the trackless ocean. A week had elapsed since he left the lordly Teneriffe, and a few days since the mountainous outline of Gomera had disappeared from sight. The memorable night was that of September 13th, 1492. There was no mistaking it; the needle of the *Santa Maria* pointed a little west of north instead of due north. Some days later, on September 17th, the pilots, having taken the sun's amplitude, reported that the variation had reached a whole point of the compass, the alarming amount of 11 degrees.

The surprise and anxiety which Columbus manifested on those occasions may be taken as indications that the phenomenon was new to him. As a matter of fact, however, his needles were not true even at the outset of the voyage from the port of Palos, where, though no one was aware of it, they pointed about 3° *east* of north. This angle diminished from day to day as the Admiral kept the prow of his caravel directed to the west, until it vanished altogether, after which the needles veered to the *west*, and kept moving westward for a time as the flag-ship proceeded on her voyage.

Columbus thus determined a place on the Atlantic in which the magnetic meridian coincided with the geographical and in which the needle stood true to the pole. Six years later, in 1498, Sebastian Cabot found another place on the same ocean, a little further north, in which the compass lay exactly in the north-and-south line. These two observations, one by Columbus and the other by Cabot, sufficed to determine the position of the *agonic line*, or line of no variation, for that locality and epoch.

The *Columbian* line acquired at once considerable importance, in the geographical and the political world, because of the proposal that was made to discard the Island of Ferro and take it for the prime meridian from which longitude would be reckoned east and west, and also because it was selected by Pope Alexander VI. to serve as a line of reference in settling the rival claims of the kingdoms of Portugal and Castile with regard to their respective discoveries. It was decided that all recently discovered lands lying to the east of that line should belong to Portugal; and those to the west, to Castile.

The line of no variation, like all other isomagnetic lines, has shifted its position with time, so that it runs to-day considerably to the west of the place assigned to it by Columbus in 1492 and by the Papal Bull of the following year.

Columbus did not speak of the disquieting observation which he made on the night of the 13th of September; he thought of it, and wondered greatly what might be the cause of such an unexpected and untoward phenomenon. His silence on the matter did not avail, for the keen-eyed sailors noticed the westerly deflection of the needle when, after a few days, it became quite apparent. They grew alarmed, believing that the laws of nature were changing as they advanced farther and farther into the unknown. It was a trying moment for the Admiral, but his ingenuity and tactfulness rose to the occasion. He told his seamen that the needle did not point to the *cynosure* or last star in the tail of the Little Bear, as commonly supposed, but to a fixed point in the celestial sphere at which there was no star, adding that the "cynosure" itself, the Polaris of our days, was not stationary, but had a rotational movement of its own like all other heavenly bodies.

We do not know what Columbus thought of his explanation, born of the stress of the moment, but the esteem in which he was held by pilots and sailors alike for his knowledge of astronomy and cosmography led them to accept it. Their fears were allayed, a mutiny was averted and a successful termination to their voyage rendered possible.

Captains of ocean-liners would give to-day a different answer to a passenger who might consult them about the splinter of steel which serves to guide their fleet vessels in darkest nights, through howling tempests and over billowy seas. The mysterious influence that controls it, they would say, comes neither from Polaris nor the pole of the world, nor from the heavens above, but from the earth beneath.

Such an explanation was not thought of until it was clearly shown a hundred years later that this globe of ours acts like a colossal lodestone, controlling every magnet in our laboratories and observatories, and every needle on board the merchantmen and fighting-monsters that plough our seas and oceans.

Without any intuition of modern theory, Columbus made two discoveries in terrestrial magnetism, as we have seen, each of fundamental importance, whether considered from the view-point of pure science or that of practical navigation, viz., (a) that the needle is not true to the pole and (b) that the angular displacement of the needle from true orientation, the *variation of the compass*, as it is called in nautical parlance, differs with the place of the observer. These two discoveries as well as the location of a place of no variation on the Atlantic Ocean entitle Columbus to a prominent place among the founders of the *science* of terrestrial magnetism.

Later observers discovered that even for a given place this element of magnetic declination has not a constant value, but undergoes changes which complete their cycle, some in a day, others in a year, and others again in centuries. The last or *secular* change in the direction of the magnetic needle was discovered by Gellibrand, of London, in 1634 (published in 1635); the *annual*, by Cassini, at Paris, 1782-1791; and the *diurnal*, by Graham, of London, in 1722.

The first observation of magnetic declination on *land* appears to have been made about the year 1510 by George Hartmann (1489-1564), Vicar of the Church of St. Sebald in Nuremberg, who found it to be 6° east in Rome, where he was living at the time. Hartmann's observation of the declination in Rome and also in Nuremberg, where the needle pointed 10° east of north, will be found in a letter which he wrote in 1544 to Duke Albert of Prussia and which remained unpublished until the year 1831.

Returning to the treatise of Peregrinus on the magnet, it should be said that for several centuries the twenty-eight manuscript copies lay undisturbed on the dusty shelves of city and university libraries. In 1562, four years after the appearance of the first printed edition (Augsburg, 1558), Taisnier, a Belgian writer on magnetics, who is also described as poet-laureate and Doctor "utriusque juris," was among the earliest to discover the "Epistola," from which he copied extensively in his little quarto on the magnet and its effects, thus showing that there were literary pirates in those days. It was also well known to Gilbert, to Cabeo and Kircher; but despite the references of these writers, the "Epistola" remained practically unknown until Cavallo, of London, called attention to the Leyden manuscript in the third edition of his "Treatise on Magnetism," 1800, by giving part of the text and accompanying it with a translation.

Later, in 1838, Libri, historian of the mathematical sciences in Italy, gave excerpts from the Paris codex with translation; but the scholar who contributed most of all to make the work of Peregrinus known is the Italian Barnabite, Timoteo Bertelli, who published in 1868 a critical study of the various manuscripts of the letter, principally those which he found in Rome and in Florence, adding copious notes of historic, bibliographic and scientific value. Father Bertelli was Professor of Physics in the Collegio della Quercia, in Florence, where he took an active interest in Italian seismology besides carrying on investigations in meteorology, telegraphy and electricity. Born in Bologna in 1826, he died in Florence in March, 1905.

The following list of manuscript copies of the "Epistola" is taken from a scholarly paper by Professor Silvanus P. Thompson, of London, which appeared in the "Proceedings of the British Academy" for 1906:—

The Bodleian Library	seven
Vatican	four

British Museum	one
Bibliothèque Nationale, Paris	two
Biblioteca Riccardiana, Florence	one
Trinity College, Dublin	one
Gonville and Caius, Cambridge	one
The University of Leyden	one
Geneva	one
Turin	one
Erfurt	three
Vienna	three
S. P. Thompson	two

The first printed edition of the "Epistola" was prepared for the press in 1558 by Achilles Gasser, a man well versed in the science and philosophy of his day; another edition, which will probably be considered the *textus receptus*, is that which was prepared and published by Bertelli in 1868.

No complete translation in any language of this historical work on magnetism was made until 1902, when Prof. Silvanus P. Thompson, of London, published his "Epistle of Peter Peregrinus of Maricourt to Sygerus of Foncaucourt, soldier, concerning the Magnet." Unfortunately, this translation was printed for private circulation and limited to 250 copies. Two years later, 1904, Brother Arnold, F. S. C., presented a memoir on Peregrinus, including a translation of the "Epistola," for the M. Sc. degree of Manhattan College, New York City, which translation was published some months later by the McGraw Publishing Company, New York. These are the only complete translations of the "Letter" of Peregrinus on the Magnet which have yet appeared.

CHAPTER II. NORMAN AND GILBERT.

We have seen that in the thirteenth century the directive property of the lodestone was recognized by Peregrinus and used by him in his pivoted compass; and that in the fifteenth, Columbus discovered magnetic declination on sea as well as its variation with place.

The next cardinal fact in terrestrial magnetism, magnetic dip, was discovered in 1576 by Robert Norman, a compass-maker of Limehouse, London. Norman possessed many of the fine qualities of mind, hand and disposition that are indispensable in the make-up of the original investigator. In pivoting his compass-needles, he soon noticed that, however carefully they were balanced before being magnetized, they did not remain horizontal after magnetization, the north-seeking end always going down through a small angle. He next had the happy idea of swinging a needle on a horizontal axis, so that it might be free to move up and down in a vertical plane, with the result that the north-seeking end again went down through a constant but much greater angle.

Like declination, the first discovered of the three magnetic elements, the dip was found to vary with place on the earth's surface, being 0° at the magnetic equator and 90° at either pole. It was with a Norman dip-circle, greatly improved, that Ross in 1831 found the north magnetic pole of the earth to be in Boothia Felix in latitude 70° 5'.3 N., and longitude 96° 45'.8 W.; and it was with a similar instrument that Amundsen recently studied the magnetic conditions of that Arctic region, the exact location of the pole itself being finally determined by an earth-inductor or spinning coil of the latest make. Though the results of his observations have not yet been made public, it is generally known that they indicate a spot for the magnetic pole close to that found by Sir James Ross. It is not expected, however, that the location of the pole by the Norwegian Commander shall exactly coincide with that of the English Captain, because the magnetic pole is believed to have nomadic tendencies of its own like our geographical pole, only much more pronounced in magnitude. After moving westward for some time at the rate of a mile per year, it retraced its steps and is now back again in the vicinity of its starting place.

Besides his dip-circle, Norman also devised a simple and very apt illustration of magnetic inclination. Thrusting a steel needle through a round piece of cork, he pared the latter down until the system, consisting of the needle and the cork, sank to a certain depth in a glass vessel containing water, and there took up a horizontal position. The needle was next removed from the water and magnetized with great care, so as not to disturb its position in the cork. When placed again in the water, the needle sank to its former depth and settled down at an angle of 71° to the horizon.

The same illustration shows another experiment which Norman made in order to determine whether the earth exerts a force of translation on a magnet, in virtue of which the magnet would tend to move bodily toward the pole. For this purpose, he floated a magnetized piece of steel wire on the surface of the water and noticed that, wherever placed, it merely swung round into the magnetic meridian without showing any tendency to move northward or southward toward the rim of the vessel. Hartmann, who observed the declination of the needle on land as stated on p. 26, appears also to have been the first to notice magnetic inclination. Having balanced a steel needle with great precision, he found that, after magnetization, it did not remain horizontal, the north-seeking end invariably dipping through an angle of 9°. The smallness of the angle in this experiment was due to the fact that the needle used by the Nuremberg Vicar could move only in a horizontal plane, whereas Norman's was free to move in a vertical circle. Had Hartmann used such a device, he would have obtained more than 60° for the dip instead of the 9° which he records.

As already remarked, the letter in which Hartmann consigns these capital observations was written in 1544, but was not published until the third decade of the nineteenth century, so that Norman has clearly the full merit of independent discovery.

In the directions which Norman gives for making observations of dip, he states explicitly that the instrument must be adjusted "duley according to the variation of the place," which means that the plane of the circle must be turned into what was called after his time "the magnetic meridian."

The discovery of magnetic dip led Norman to discard the view generally held in his time, which placed the controlling influence of the compass-needle in far-off celestial space; for he says that the *poynt respective* which the magnet indicates, but to which it is not bodily drawn, is not in the heavens above, but in the earth itself. His words are: "And by the declining of the needle is also proved that the poynt respective is rather in the earth than in the heavens, as some have imagined; and the greatest reason why they so thought, as I judge, was because they were never acquainted with this declining in the needle."

Here we have a radical departure from the scientific creed of the time, a notable advance in scientific theory, an entirely new philosophy founded by Norman, the compass-maker, and greatly developed twenty-four years later by his fellow-citizen, Gilbert, the physician.

Norman made another remark of great importance in the new philosophy, the justness of which was appreciated by Gilbert, his contemporary, but more so by Faraday and Clerk Maxwell, two centuries later. It refers to the space surrounding a magnet, natural or artificial, which cubical space Gilbert, following Norman, called an *orb of virtue*. That the influence or "effluvium" of the magnet extends throughout the entire space may readily be seen by carrying a compass-needle round a magnet from point to point, far away as well as close by. The phrase "orb of virtue," or sphere of magnetic influence, appears to describe the actual magnetic condition of the space in question more pertinently than our modern equivalent of "magnetic field."

The words of Norman are very remarkable: "I am of opinion that if this vertue could by anie means be made visible to the eie of man, it would be found in a sphericall forme, extending round about the stone in great compasse and the dead bodie of the stone in the middle thereof." The lines which immediately follow this statement, pregnant with significance, show the deep religious feeling of the author. They read: "and this I have partly proved and made visible to be seene in some manner, and God sparing mee life, I will herein make further experience and that not curiouslie but in the feare of God as neere as He shall give me grace and meane to annexe the same unto a booke of navigation which I have had long in hand."—Chap. VIII.

It is evident from the pages of the *Newe Attractive* (1581) that Norman was animated with the right spirit of inquiry, which is calm, deliberate and judicious, which leads to the discovery of facts, to their coordination and experimental illustration before explanations are thought of and long before new theories are propounded. The style in which this little treatise is written has a charm of its own, mainly by reason of its quaintness. At the end of his address to the candid reader, which, after the manner of the times, was somewhat belabored and rhetorical in character, Norman breaks away from common inadequate prose; and, giving wings to his imagination, writes a lyric on the magnet which is the first metrical composition in English that we have on such a subject. It reads:—

THE MAGNES OR LOADSTONE'S CHALLENGE.
Give place ye glittering sparks, ye glimmering Diamonds bright, Ye Rubies red, and Saphires brave wherein ye most delight.
In breefe, yee stones inricht, and burnisht all with golde, Set forth in Lapidaries shops, for Jewells to be sold.
Give place, give place I say, your beautie, gleame and glee, Is all the vertue for the which, accepted so you bee.
Magnes, the Loadstone I, your painted sheath defie, Without my help in Indian seas, the best of you might lie.
I guide the Pilot's course, his helping hand I am, The Mariner delights in me, so doth the Marchant man.
My vertue lies unknowne, my secrets hidden are, By me, the Court and Commonweale, are pleasured very farre.

No ship could sail on Seas, her course to run aright, Nor Compass shew the ready way were Magnes not of might.
Blush then, and blemish all, bequeath to mee thats due, Your seats in golde, your price in plate, which Jewellers do renue.
Its I, its I alone, whom you usurp upon, Magnes my name, the Loadstone cal'd, the prince of stones alone.
If this you can deny, then seem to make reply, And let the painfull sea-man judge, the which of us doth lie.
The Mariner's Judgement.
The Loadstone is the stone, the onely stone alone, Deserving praise above the rest whose vertues are unknown.
The Marchant's Verdict.
The Diamonds bright, the Saphires brave, Are stones that bear the name, but flatter not, and tell the troath, Magnes deserves the same.
(Edition of 1720.)

Norman's *Newe Attractive* was well known to Gilbert, as were also the *Epistola* of Peregrinus, the *Magiae Naturalis* of Porta, and indeed all books treating of the lodestone, the magnet, or the compass-needle. His own work *De Magnete*, published in the year 1600, is a compendium of the world's knowledge of magnetism and electricity at the time. In its pages, he not only discusses the opinions of others, but describes discoveries of his own made during the twenty years which he ardently devoted to the pursuit of experimental science, crowning his investigations with theories in electricity and magnetism as became a true philosopher.

Impressed by the originality of Gilbert's treatise, the practical ingenuity and philosophic acumen displayed throughout, Hallam wrote in his *Introduction to the Literature of Europe*: "Gilbert not only collected all the knowledge which others had possessed on the subject, but became at once the father of experimental philosophy in this island; and, by a singular felicity and acuteness of genius, the founder of theories which have been received after the lapse of ages and are almost universally received into the creed of science."

At a period when natural science was taught in the schools of Europe mainly from text-books, we find Gilbert proclaiming by example and advocacy the paramount value of experiment for the advancement of learning. He was unsparing in his denunciation of the superficiality and verbosity of mere bookmen, and had no patience with writers who treated their subjects "esoterically, reconditely and mystically." For him, the laboratory method was the only one that could secure fruitful results and contribute effectively to the advancement of learning.

It is true that men of unusual ability and strong character strove before his time to adjust the claims of authority in matters scientific. While respectful of the teachings of recognized leaders, they were not, however, awed into acquiescence by an academical "magister dixit." On the contrary, they wanted to test with their eyes in order to judge with reason; believing in the importance of experiment, they sought to acquire a knowledge of nature from nature herself.

Such were Albert the Great and Friar Bacon. Albert did not bow obsequiously to the authority of Aristotle or any of his Arabian commentators; he investigated for himself and became, for his age, a distinguished botanist, physiologist and mineralogist.

The Franciscan monk of Ilchester has left us in his *Opus Majus* a lasting memorial of his practical genius. In the section entitled "Scientia Experimentalis," he affirms that "Without experiment, nothing can be adequately known. An argument proves theoretically, but does not give the certitude necessary to remove all doubt, nor will the mind repose in the clear view of truth, unless it find it by way of experiment." And in his *Opus Tertium*: "The strongest arguments prove nothing, so long as the conclusions are not verified by experience. Experimental science is the queen of sciences and the goal of all speculation."

No one, even in our own times, wrote more strongly in favor of the practical method than did this follower of St. Francis in the thirteenth century. Being convinced that there can be no conflict between scientific and revealed truths, he became an irrepressible advocate for observation and experiment in the study of the phenomena and forces of nature.

The example of Peregrinus, of Albert and Friar Bacon, not to mention others like Vincent of Beauvais, the Dominican encyclopedist, was, however, not sufficient to wean students from the easy-going routine of book-learning. A few centuries had to elapse before the weaning was effectively begun; and the man who contributed in a marked degree to this result was Gilbert the Philosopher of Colchester (1544-1603).

Having received the elements of his education in the Grammar School of Colchester, his native town, Gilbert entered St. John's College, Cambridge, from which university he took his B. A. degree in 1560, M. A. in 1564 and M. D. in 1569. In all, he appears to have been connected with the University for a period of eleven or twelve years, as student, Fellow, and examiner.

On leaving Cambridge, Gilbert traveled for four years on the Continent, principally in Italy, visiting medical schools and studying methods of treatment under the leading physicians and surgeons of the day as well as discussing scientific theory with the leaders of thought. On his return to England in 1573, he practised medicine in London "with great applause and success." He was elected President of the Royal College of Physicians in 1599, and appointed Physician to Queen Elizabeth in 1601 and to her successor, James I., in 1603.

On one occasion, he hears that Baptista Porta, whom he calls "a philosopher of no ordinary note," said that a piece of iron rubbed with a diamond turns to the north. He suspects this to be heresy. So, forthwith he proceeds to test the statement by experiment. He was not dazzled by the reputation of Baptista Porta; he respected Porta, but respected truth even more. He tells us that he experimented with seventy diamonds in presence of many witnesses, employing a number of iron bars and pieces of wire, manipulating them with the greatest care while they floated on corks; and concludes his long and exhaustive research by plaintively saying: "Yet never was it granted me to see the effect mentioned by Porta."

Though it led to a negative result, this probing inquiry was a masterpiece of experimental work.

Gilbert incidentally regrets that the men of his time "are deplorably ignorant with respect to natural things," and the only way he sees to remedy this is to make them "quit the sort of learning that comes only from books and that rests only on vain arguments and conjectures," for he shrewdly remarks that "even men of acute intelligence without actual knowledge of facts and in the absence of experiment easily fall into error."

Acting on this intimate conviction, he labored for twenty years over the theories and experiments which he sets forth in his great work on the magnet. "There is naught in these books," he tells us, "that has not been investigated, and again and again done and repeated under our eyes." He begs any one that should feel disposed to challenge his results to repeat the experiments for himself "carefully, skilfully and deftly, but not heedlessly and bunglingly."

It has been said that we are indebted to Sir Francis Bacon, Queen Elizabeth's Chancellor, for the inductive method of studying the phenomena of nature. Bacon's merit lies in the fact that he not only minutely analyzed the method, pointing out its uses and abuses, but also that he showed it to be the only one by which we can attain an accurate knowledge of the physical world around us. His sententious eulogy went forth to the world of scholars invested with all the importance, authority and dignity which the high position and worldwide fame of the philosophic Chancellor could give it. But while Bacon thought and wrote in his study, Gilbert labored and toiled in his workshop. By his pen, Bacon made a profound impression on the philosophic mind of his age; by his researches, Gilbert explored two provinces of nature and added them to the domain of science. Bacon was a theorist, Gilbert an investigator. For twenty years he shunned the glare of society and the throbbing excitement of public life; he wrenched himself away from all but the strictest exigencies of his profession, in order to devote himself undistractedly to the pursuit of science. And all this forty years before the appearance of Bacon's *Novum Organum*, the very work which contains the philosopher's "large thoughts and lofty phrases" on the value of experiment as a means for the advancement of learning. During that long period Gilbert haunted Colchester, where he delved into the secrets of nature and prepared the materials for his great work on the magnet. The publication of this Latin treatise made him known in the universities at home and especially abroad: he was appreciated by all the great physicists and mathematicians of his age; by such men as Sir Kenelm Digby; by William Barlowe, a great "magneticall" man; by Kepler, the astronomer, who adopted and defended his views; by Galileo himself, who said: "I extremely admire and envy the author of *De Magnete*."

The science of magnetism owes more to Gilbert than to any other man, Peregrinus (1269) excepted. He repeated for himself the numerous and ingenious experiments of the medieval philosopher, and added much of his own which he discovered during the long period of a life devoted to the diligent exploration of this domain in the world of natural knowledge.

The ancients spoke of the lodestone as the Magnesian stone, from its being found in abundance in the vicinity of Magnesia, a city of Asia Minor. In his Latin treatise of 254 (small) folio pages, Gilbert uses the adjective form of the term, but never the noun "Magnetismus" itself. Our English term *magnetism* appears for the first time on page 2 of Archdeacon Barlowe's "Magneticall Advertisements," published in 1616; while the surprising compound, "electro-magnetismos," is the title of a chapter in Father Kircher's "Magnes, sive de Arte Magnetica," printed in the year 1641.

Gilbert showed that a great number of bodies could be electrified; but maintained that those only could exhibit magnetic properties which contain iron. He satisfies himself of this by rubbing with a lodestone such substances as wood, gold, silver, copper, zinc, lead, glass, etc., and then floating them on corks, quaintly adding that they show "no poles, because the energy of the lodestone has no entrance into their interior."

To-day we know that nickel and cobalt behave like iron, whilst antimony, bismuth, copper, silver and gold are susceptible of being influenced by powerful electro-magnets, showing what has been termed diamagnetic phenomena. Even liquids and gases, in Faraday's classical experiments, yielded to the influence of his great magnet; and Professor Dewar, in the same Royal Institution, exposed some of his liquid air and liquid oxygen to the influence of Faraday's electromagnet and found them to be strongly attracted, thus behaving like the paramagnetic bodies, iron, nickel and cobalt.

Gilbert observes in all his magnets two points, one near each end, in which the force, or, as he terms it, "the supreme attractional power," is concentrated. Like Peregrinus, he calls these points the *poles* of the magnet, and the line joining them its magnetic axis. With the aid of his steel versorium, he recognizes that similar poles are mutually hostile, whilst opposite poles seize and hold each other in friendly embrace. He also satisfies himself that the energy of magnets resides not only in their extremities, but that it permeates "their inmost parts, being entire in the whole and entire in each part." This is exactly what Peregrinus said in 1269 and what we say to-day; it is nothing else than the molecular theory proposed by Weber, extended by Ewing and universally accepted.

At any rate, Gilbert is quite certain that whatever magnetism may be, it is not, like electricity, a material, ponderable substance. He ascertained this by weighing in the most accurate scales of a goldsmith a rod of iron before and after it had been rubbed with the lodestone, and then observing that the weight is precisely the same in both cases, being "neither less nor more."

Without referring to the prior discovery of Norman, whom he calls "a skilled navigator and ingenious artificer," Gilbert satisfies himself that not only the magnet, but all the space surrounding it, possesses magnetic properties; for the magnet "sends its force abroad in all directions, according to its energy and quality." This region of influence Norman called a sphere of "vertue," and Gilbert an "orbis virtutis," which is the Latin equivalent; we call it a "magnetic field," or field of force, which is less expressive and less appropriate. With wonderful intuition, Gilbert sees this space filled with lines of magnetic virtue passing out radially from his spherical lodestone, which lines he calls "rays of magnetic force."

Clerk Maxwell was so fascinated with this beautiful concept that he made it the work of his life to study the field of force due to electrified bodies, to magnets and to conductors conveying currents; his powerful intellect visualized those lines and gave them accurate mathematical expression in the great treatise on electricity and magnetism which he gave to the world in 1873.

Gilbert observes that the lodestone may be spherical or oblong; "whatever the shape, imperfect or irregular, verticity is present; there are poles," and the lodestones "have the selfsame way of turning to the poles of the world." He knows that a compass-needle is not drawn bodily towards the pole, and does not hesitate in this instance to give credit to his countryman, Robert Norman, for having clearly stated this fact and aptly demonstrated it. Following Norman, he floats a needle in a vessel by means of a piece of cork, and notices that on whatever part of the surface of the water it may be placed, the needle settles down after a few swings invariably in the same direction. His words are: "It revolves on its iron center and is not borne towards the rim of the vessel."

Gilbert knew nothing about the mechanical couple that came into play, but he knew the fact; and, with the instinct of the philosopher, tested it in a variety of ways.

We explain the orientation of the compass-needle by saying that it is acted upon by a pair of equal and opposite forces due to the influence of the terrestrial magnetic poles on each end of the needle and by showing that such a couple can produce rotation, but not translation.

We find Gilbert working not only with steel needles and iron bars, but also with rings of iron. He strokes them with a natural magnet and feels certain that he has magnetized them. He assures us that "one of the poles will be at the point rubbed and the other will be at the opposite side." To show that the ring is really magnetized, he cuts it across, opens it out, and finds that the ends exhibit polar properties.

A favorite piece of apparatus with Gilbert, as with Peregrinus, was a lodestone ground down into globular form. He called it a terrella, a miniature earth, and used it extensively for reproducing the phenomena described by magnetizers, travelers and navigators. He breaks up terrellas, in order to examine the magnetic condition of their inner parts. There is not a doubtful utterance in his description of what he finds; he speaks clearly and emphatically. "If magnetic bodies be divided, or in any way broken up, each several part hath a north and a south end"; *i.e.*, each part will be a complete magnet.

We find him also comparing magnets by what is known to us as the "magnetometer method." He brings the magnetized bars in turn near a compass-needle and concludes that the magnet or the lodestone which is able to make the needle go round is the best and strongest. He also seeks to compare magnets by a process of weighing, similar to what is called, in laboratory parlance, the "test-nail" method. He also inquires into the effect of heat upon his magnets, and finds that 'a lodestone subjected to any great heat loses some of its energy.' He applies a red-hot iron to a compass-needle and notices that it 'stands still, not turning to the iron.' He thrusts a magnetized bar into the fire until it is red-hot and shows that it has lost all magnetic power. He does not stop at this remarkable discovery, for he proceeds to let his red-hot bars cool while lying in various positions, and finds: (1) that the bar will acquire magnetic properties if it lie in the magnetic meridian; and (2) that it will acquire none if it lie east and west. These effects he rightly attributes to the inductive action of the earth.

Gilbert marks these and other experiments with marginal asterisks; small stars denoting minor and large ones important discoveries of his. There are in all 21 large and 178 small asterisks, as well as 84 illustrations in *De Magnete*. This implies a vast amount of original work, and forms no small contribution to the foundations of electric and magnetic science.

Gilbert clearly realized the phenomena and laws of magnetic induction. He tells us that "as soon as a bar of iron comes within the lodestone's sphere of influence, though it be at some distance from the lodestone itself, the iron changes instantly and has its form renewed; it was before dormant and inert; but now is quick and active." He hangs a nail from a lodestone; a second nail from the first, a third from the second and so on—a well-known experiment, made every day for elementary classes. Nor is this all, for he interposes between the lodestone and his iron nail, thick boards, walls of pottery and marble, and even metals, and he finds that there is naught so solid as to do away with its force or to check it, save a plate of iron. All that can be added to this pregnant observation is that the plate of iron must be very thick in order to carry all the lines of force due to the magnet, and thus completely screen the space beyond.

But Gilbert is astonishing when he goes on to make thick boxes of gold, glass and marble; and, suspending his needle within them, declares with excusable enthusiasm that, regardless of the box which imprisons the magnet, it turns to its predestined points of north and south. He even constructs a box of iron, places his magnet within, observes its behavior, and concludes that it turns north and south, and would do so were "it shut up in iron vaults sufficiently roomy." In this, he was in error, for experiments show that if the sides of the box are thin, the needle will experience the directive force of the earth; but if they are sufficiently thick—thick as the walls of an ordinary safe—the inside of such a box will be completely screened; none of the earth's magnetic lines will get into it so that the needle will remain indifferently in any position in which it is placed. Some years ago, the physical laboratory of St. John's College, Oxford, was screened from the obtrusive lines of neighboring dynamos by building two brick walls parallel to each other and eight inches apart and filling in the space with scrap iron. A delicate magnetometer showed that such a structure allowed no leakage of lines of force through it, but offered an impenetrable barrier to the magnetic influence of the working dynamos.

Gilbert's greatest discovery is that the earth itself acts as a vast globular magnet having its magnetic poles, axis and equator. The pole which is in our hemisphere, he variously calls north, boreal or arctic. Whilst that in the other hemisphere he calls south, austral or antarctic. He sought to explain the magnetic condition of our globe by the presence, especially in its innermost parts, of what he calls true, terrene matter, homogeneous in structure and endowed with magnetic properties, so that every separate fragment exhibits the whole force of magnetic matter. He is quite aware that his theory is a grand generalization; and admits that it is "a new and till now unheard-of view," and so confident is he in its worth that he is not afraid to say that "it will stand as firm as aught that ever was produced in philosophy, backed by ingenious argumentation or buttressed by mathematical demonstration."

In developing his theory of terrestrial magnetism, Gilbert fell into certain errors, chiefly for want of data, but partly also by reason of his adherence to the view that the earth exactly resembled his terrella in its magnetic action. Accordingly, he believed that the magnetic poles of the earth were diametrically opposite each other and that they coincided with the poles of rotation, whence it followed that the magnetic meridian everywhere coincided with the geographical, and that the magnet, unless influenced by local disturbances, stood true to the pole.

It was, however, well known from the thrilling experience of Columbus and the constant report of travelers that this was not the case. Gilbert himself says that at the time of writing, in the year 1600, the needle pointed 11-1/3° east of north in London; but what he did not know and could not have known was that this easterly deviation was decreasing from year to year, to vanish altogether in 1657, after which the needle began to decline to the west.

This magnetic declination sorely perplexed Gilbert, as it did not fit in with his theory. Yet an explanation was needed; and as the earth must be considered a normal and well-behaved magnet, though of cosmical size, Gilbert turns the difficulty by saying that this variation is nothing else than "a sort of perturbation of the directive force" caused by inequalities in the earth's surface by continents and mountain masses: "Since the earth's surface is diversified by elevations of land and depths of seas, great continental lands, oceans and seas differing in every way while the power that produces all magnetic movements comes from the constant magnetic earth-substance which is strongest in the most massive continent and not where the surface is water or fluid or unsettled, it follows that toward a massive body of land or continent rising to some height in any meridian, there is a measurable magnetic leaning from the true pole toward the east or the west."

So convinced is Gilbert of the true and satisfactory character of his explanation that he goes on to say that, "In northern regions, the compass varies because of the northern eminences; in southern regions, because of southern eminences. On the equator, if the eminences on both sides were equal, there would be no variation." In a later chapter of Book IV., he adds that, "in the heart of great continents there is no variation; so, too, in the midst of great seas."

As continents and mountain-chains are among the permanent features of our planet, Gilbert concluded that the misdirection of the needle was likewise permanent or constant at any given place, a conclusion which observations made after Gilbert's time showed to be incorrect. Gilbert writes: "As the needle hath ever inclined toward the east or toward the west, so even now does the arc of variation continue to be the same in whatever place or region, be it sea or continent; so, too, will it be forever unchanging."

This we know to be untrue, and Gilbert, too, could have known as much had he brought the experimental method, which he used with such consummate skill and fruitful results in other departments of his favorite studies, to bear on this particular element of terrestrial magnetism. He labored with incredible ardor and persistence for twenty years in his workshops at Colchester over the experiments in electricity, magnetism and terrestrial magnetism which he embodies and discusses in his original and epoch-making book, *De Magnete*, published in the year 1600; a period of twenty years was long enough for such a careful observer as he was to detect the slow change in magnetic declination discovered by his friend Gellibrand in 1634, published by him in 1635, and known to-day as the "secular variation." It is true the quantity to be measured was small; but what is surprising is that such an industrious and resourceful experimenter as Gilbert was does not record in his pages any observations of his own on declination or dip, elements of primary importance in magnetic theory.

Shortly after the voyage of Columbus it was thought that the *longitude* of a place could be found from its magnetic declination. Gilbert, however, did not think so, and accordingly scores those who championed that view. "Porta," he says, "is deluded by a vain hope and a baseless theory"; Livius Sanutus "sorely tortures himself and his readers with like vanities"; and even the researches of Stevin, the great Flemish mathematician, on the cause of variation in the southern regions of the earth are "utterly vain and absurd."

With regard to dip, Gilbert erroneously held that for any given latitude it had a constant value. He was so charmed with this constancy that he proposed it as a means of determining *latitude*. There is no diffidence in his mind about the matter; he is sure that with his "inclinatorium" or dip-circle, together with accompanying tables, calculated for him by Briggs, of logarithmic fame, an observer can find his latitude "in any part of the world without the aid of the sun, planets or fixed stars in foggy weather as well as in darkness."

After such a statement, it is no wonder that he waxes warm over the capabilities of his instrument and allows himself to exclaim: "We can see how far from idle is magnetic philosophy; on the contrary, how delightful it is, how beneficial, how divine! Seamen tossed by the waves and vexed with incessant storms while they cannot learn even from the heavenly luminaries aught as to where on earth they are, may with the greatest ease gain comfort from an insignificant instrument and ascertain the latitude of the place where they happen to be."

Gilbert dwells at length on the inductive action of the earth. He hammers heated bars of iron on his anvil and then allows them to cool while lying in the magnetic meridian. He notes that they become magnetized, and does not fail to point out the polarity of each end. He likewise attributes to the influence of the earth the magnetic condition acquired by iron bars that have for a long time lain fixed in the north-and-south position and ingenuously adds: "for great is the effect of long-continued direction of a body towards the poles." To the same cause, he attributes the magnetization of iron crosses attached to steeples, towers, etc., and does not hesitate to say that the foot of the cross always acquires north-seeking polarity.

In a similar manner, every vertical piece of iron, like railings, lamp-posts, and fire-irons, becomes a magnet under the inductive action of the earth. In the case of our modern ships, the magnetization of every plate and vertical post, intensified by the hammering during construction, converts the whole vessel into a magnetic magazine, the resulting complex "field" rendering the adjustment of the compasses somewhat difficult and unreliable. The unreliable character of the adjustment arises mainly from the changing magnetism of the ship with change of place in the earth's magnetic field, the effect increasing slightly from the magnetic equator to the poles.

With luminous insight into the phenomena of terrestrial magnetism, Gilbert observes that in the neighborhood of the poles, a compass-needle, tending as it does to dip greatly, must in consequence experience only a feeble directive power. To which he adds that "at the poles there is no direction," meaning, no doubt, that a compass-needle would remain in any horizontal position in which it might be placed when in the vicinity of the magnetic pole.

This is precisely the experience of all Arctic explorers, who find that their compasses become less and less active as they sail northward, the reason being that the horizontal component of the earth's magnetic force, which alone controls the movements of the compass-needle, decreases as the ship advances and vanishes altogether at the magnetic pole. When once a high latitude is reached, captains do not depend upon their compasses for their bearings, but have recourse to astronomical observations. In his account of magnetic work carried on in the neighborhood of the magnetic pole, Amundsen says: "At Prescott Island the compass, which for some time had been somewhat sluggish, refused entirely to act, and we could as well have used a stick to steer by."

As a physician, Gilbert valued iron for its medicinal properties, but denounced quacks and wandering mountebanks who practised "the vilest imposture for lucre's sake," using powdered lodestone for the cure of wounds and disorders. "Headaches," he said, "are no more cured by application of a lodestone than by putting on an iron helmet or a steel hat"; and again: "To give it in a draught to dropsical persons is either an error of the ancients or an impudent tale of their copyists." Elsewhere he condemns prescriptions of lodestone as "an evil and deadly advice" and as "an abominable imposture."

In the sixth and last book of *De Magnete*, Gilbert sets forth his views on such astronomical subjects as the figure of the earth, its suspension in space, rotation on its axis and revolution around the sun.

As to the figure of our planet, the primitive view widely credited in early times was that the earth is a flat, uneven mass floating in a boundless ocean. The Hindoos, however, did not accept the flatland doctrine, but taught that the earth was a convex mass which rested on the back of a triad of elephants having for their support the carapace of a gigantic tortoise. Of course, they did not say how the complaisant chelonian contrived to maintain his wonderful state of equilibrium under the superincumbent mass.

Aristotle (384-322, B. C.) taught that the earth, fixed in the center of the universe, is not flat as a disk, but round as an orange, giving as proofs (1) the gradual disappearance of a ship standing out to sea and (2) the form of the shadow cast by the earth in lunar eclipses, to which others added (3) the change in the altitude of circumpolar stars readily noticeable in traveling north or south. Aristarchus of Samos (310-250), one of the great astronomers of antiquity, went further, not fearing to teach that the earth is spherical in form, that it turns on its axis daily and revolves annually around the sun. Such orthodox teaching did not, however, commend itself to people generally, as they did not exactly like the idea of being whisked round with their houses and cities at a dangerous speed, preferring to explain celestial phenomena by the rotation of the vast celestial sphere, with all the starry host, round a flat, immovable earth. For them such a system of cosmography recommended itself by its simplicity and reasonableness as well as by the sense of stability, rest and comfort which it brought along with it.

Ptolemy, who flourished at Alexandria about 150 A. D. and whose name is associated with a system of the world, also held that the earth is spherical in form, giving at the same time some very ingenious proofs of his belief. St. Augustine, in the fourth century, was not opposed to the doctrine of a round earth, though he felt the religious difficulty arising from the existence of the antipodes, which difficulty reached its acute stage four hundred years later.

It is well to remember that the Church did not condemn the existence of an antipodean world; what it did condemn was the teaching of Virgilius, Bishop of Salzburg, to the effect that this world, lying under the equator, was inhabited by a race of men not descended from Adam. Virgilius also taught that the antipodes had a sun and moon different from ours, an astronomical opinion for which he was never molested by ecclesiastical authority.

Boethius, the worthy representative of the natural and the higher philosophy of the sixth century, wrote of the earth as globe-like in form, but small in comparison with the heavens. Isidore of Seville, in the seventh century, "the most learned man of his age," and the encyclopedic Bede, in the eighth, rejected the theory of a flat, discoidal earth and returned to the spherical form of the early Greek astronomers. But again, centuries had to elapse before people could be brought to tolerate views of the world that seemed so directly opposed to the daily testimony of their senses.

The strong, conclusive arguments which alone establish this theory on a firm basis were, however, not known to Copernicus and could not have been known in an age that preceded the invention of the telescope and in which the astronomer had to be the constructor of his own crude wooden instruments. The wonder is that Copernicus did such excellent observational work on the banks of the Vistula with the rough appliances at his disposal. The arguments which he put forward and urged with consummate skill for the acceptance of his revolutionary theory were its general simplicity and probability. Of proofs clear and decisive, he gave none; yet, while he was working on his epoch-making treatise, begun in 1507 and published in 1543 with dedication to Pope Paul III., a direct proof of the earth's spherical form was given by the return (1528) from the Philippines along an eastern route of one of Magellan's ships, which had reached those distant isles after crossing the western ocean, which the Portuguese navigator called the "Pacific," from the tranquility of its waters. For a direct proof of the earth's *annual* motion, the world had to wait two hundred years more, until Bradley discovered the "aberration of light" in 1729; and for a direct demonstration of its *diurnal* motion until Foucault made his pendulum experiment in the Panthéon, in 1851.

We cannot let Gilbert's reference to a "weightless earth" pass without a few remarks to justify our approval of the statement.

The idea connoted by the term weight is the pull which the earth exerts on the mass of a body; thus, when we say that an iron ball weighs six pounds, we mean that the earth pulls it downwards with a force equal to the weight of six pounds. That the weight of a given lump of matter is not a constant but a dependent quantity may be seen from a number of considerations. Its weight in vacuo, for instance, is different from its weight in air, and this latter differs considerably from its weight in water or in oil. Again, if we take our experimental ball down the shaft of a mine, the spring-balance used to measure the pull of the earth on it will not record six pounds but something less; and the further we descend, the less will the spring-balance be found to register. At a depth of two thousand miles below the surface, the ball would be found to have lost half its weight; and at a depth of four thousand, all its weight. At the earth's center a "box of weights" would still be called a "box of weights," though neither the box itself nor its enclosed standards singly or collectively would have any weight whatever. It has been shown experimentally that two masses weigh slightly less when placed one above the other than when placed side by side, because in the latter case their common mass-center is measurably nearer to the center of the earth. Every mother knows that when a boy is sent to buy a pound of candy, it is the mass of the sweet stuff that makes him happy, and not its weight, for this acts more like an incumbrance while he is bringing it home. Of course, weight is every day used, and correctly, as a measure of mass, for every student of mechanics writes without the least hesitation,

$$W=Mg.$$

by which he simply means that the weight of a body is directly proportional to its mass (M), which is constant wherever the body may be taken, and to the intensity of gravity (g), which varies slightly with geographical position. As both scale-pans of an ordinary balance are equally affected by the local value of *g*, it follows that equilibrium is established only when the two *masses*—that of the body and that of the standards—are themselves equal: hence weighing is in reality only a process of comparing masses, *i.e.*, a process of "massing."

If we bring our experimental ball to the top of a hill or to the summit of a mountain or aloft in a balloon, we find the pull on the registering spring growing less and less as we go higher and higher, from which we naturally conclude that if we could go far enough out into circumterrestrial space, say, towards the moon, the ball would lose its weight entirely; it would cease to stretch the spring of the measuring balance, its weight vanishing at a definite, calculable distance from the earth's center. If carried beyond that point the ball would come under the moon's preponderating attraction and would begin to depress anew the index of the balance until at the surface of our satellite it would be found to weigh exactly *one* pound. If transferred to the planet Mars the ball would weigh *two* pounds, and if to the surface of the giant planet Jupiter, *sixteen* pounds. But while its weight thus changes continually, its mass or quantity of matter, the stuff of which it is made, remains constant all the while, being equally unaffected by such variables as motion, position or even temperature.

Returning from celestial space to our more congenial terrestrial surroundings, we find a similar inconstancy in the weight of the ball as we travel from the equator toward either pole, the weight being least at the equator and slightly greater at either end of our axis of rotation. This change is fully accounted for by the spheroidal figure of the earth and its motion of rotation, in virtue of which, while going from the equator toward the pole, our distance from the center of attraction undergoes a slight diminution, as does also the component of the local centrifugal force, which is in opposition to gravity.

From all this, it will be seen that the weight of a body is more of the nature of an accidental rather than an essential property of matter, whereas its mass is a necessary and unvarying property. Hence we speak with propriety of the conservation of mass just as we speak with equal propriety of the conservation of energy; but we may never speak or write of the conservation of weight. The mass of our iron ball is precisely the same away from the surface of the earth as it is anywhere on the surface, whether a thousand miles below the surface or a thousand miles above it; and the same it would be found in any part of the solar system or of the starry universe to which it might be taken.

Since weight is nothing else than the pull which the earth exerts on a body, it follows that, big and massive as our planet is, it must, nevertheless, be weightless; for it cannot with any degree of propriety be said to pull itself. It is incapable of producing even an infinitesimal change in the position of its mass-center, or center of gravity, as this centroid is sometimes called. The earth attracts *itself* with no force whatever; but is attracted and governed in its annual movement by the sun, the central controlling body of our system, while the moon and planets play only the part of petty disturbers.

It would, however, be right to speak of the *weight* of the earth *relatively* to the sun; for the sun attracts the mass of our planet with a certain definite force, readily calculable from the familiar formula for central force, viz., mv^2/r., in which m is the mass of the earth, v its orbital velocity and r its distance from the sun. Supplying the numbers, the weight of the earth relatively to the sun, comes out to be

3,000000,000000,000000 or 3×10^{18} tons weight,

or, in words, three million million million tons weight.

It may here be noted that the velocity v of the earth in its orbit is a varying quantity, depending on distance from the sun. As this distance is least in December and greatest in June, it follows that the earth is heavier relatively to the sun in winter than it is in summer.

The *mass* of the earth, on the other hand, is not a relative and variable quantity, but a constant and independent one, which would not be affected either by the sudden annihilation of all the other members of the solar system or by the instantaneous or successive addition of a thousand orbs. Mass being the product of volume by density, that of the earth is 6000,000000,000000,000000 or 6×10^{21} tons mass, which reads six thousand million million million tons mass.

The number which expresses the mass of the earth is thus very different from that which represents its weight relatively to the sun. It is obvious that the latter would be a much greater quantity if our planet were transferred to the orbit of Venus and very much less if transferred to that of far-off Jupiter, but the number which expresses its *mass* would remain precisely the same in both cases, viz., the value given above.

In elaborating his theory of magnetism, and especially his magnetic theory of the earth, Gilbert made extensive use of lodestone-globes, which he called "terrellas," *i.e.,* miniature models of the earth. In pursuing his searching inquiry, he was gradually led from these "terrellas" to his great induction that the earth itself is a colossal, globe-like magnet. Following Norman, "the ingenious artificer," of Limehouse, London, he also showed that the entire cubical space which surrounds a lodestone is an "orb of virtue," or region of influence, from which he inferred that the earth itself must have its "orb of virtue," or magnetic field, extending outward to a very great distance.

Gilbert does not, for a moment, think that this theory of terrestrial magnetism, the first ever given to the world, is a wild speculation. Far from it; he is convinced that "it will stand as firm as aught that ever was produced in philosophy, backed by ingenious argumentation or buttressed by mathematical demonstration."

If the earth has a magnetic field, he argued, why not the moon, the planets and the sun itself, "the mover and inciter of the universe"? Given these planetary magnetic fields, Gilbert seems to have no difficulty in finding out the forces necessary to account for the crucial difficulties of the Copernican doctrine. Nor is the medium absent that is needed for the mutual action of magnetic globes, for we are assured that it is none other than the universal *ether*, which, he says "is without resistance."

Gilbert disposes of the cosmographic puzzle of the "suspension" of the earth in space by saying, and saying justly, that the earth "has no heaviness of its own," and, therefore, "does not stray away into every region of the sky." To emphasize the statement, he continues: "The earth, in its own place, is in no wise heavy, nor does it need any balancing"; and again, "The whole earth itself has no weight." "By the wonderful wisdom of the Creator," he elsewhere says, "forces were implanted in the earth that the globe itself might with steadfastness take direction."

Gilbert holds that the daily rotation of the earth on its axis is also caused, and maintained with strict uniformity, by the same prevalent system of magnetic forces, for "lest the earth should in divers ways perish and be destroyed, she rotates in virtue of her *magnetic energy*, and such also are the movements of the rest of the planets."

Just how this magnetic energy acts to produce the rotatory motion of a massive globe Gilbert does not say. Nor was he able to solve such a magnetic riddle, for there was nothing in his philosophy to explain how a lodestone-globe in free space should ever become a perpetual magnetic motor. Oddly enough he disagrees with Peregrinus, who maintained in his *Epistola*, 1269, that a terrella, or spherical lodestone, poised in the meridian, would turn on its axis regularly every 24 hours. He naively says: "We have never chanced to see this; nay, we doubt if there is such a movement." Continuing, he brings out his clinching argument: "This daily rotation seems to some philosophers wonderful and incredible because of the ingrained belief that the mighty mass of earth makes an orbital movement in 24 hours; it were more incredible that the moon should in the space of 24 hours traverse her orbit or complete her course; more incredible that the sun and Mars should do so; still more that Jupiter and Saturn; more than wonderful would be the velocity of the fixed stars and firmament."

Here he finds himself obliged to berate Ptolemy for being "over-timid and scrupulous in apprehending a break up of this nether world were the earth to move in a circle. Why does he not apprehend universal ruin, dissolution, confusion, conflagration and stupendous celestial and super-celestial calamities from a motion (that of the starry sphere) which surpasses all imagination, all dreams and fables and poetic license, a motion ineffable and inconceivable?"

Gilbert is not clear and emphatic on the other doctrine of Copernicus, the revolution of the earth and planets around the sun. He does, however, say that each of the moving globes "has circular motion either in a great circular orbit or on its own axis, or in both ways." Again: "The earth by some great necessity, even by a virtue innate, evident and conspicuous, is turned circularly about the sun." Elsewhere he affirms that the moon circles round the earth "by a magnetic compact of both." He returns to this point in his *De Mundo*

Nostro, saying, "The force which emanates from the moon reaches to the earth; and, in like manner, the *magnetic virtue* of the earth pervades the region of the moon."

We have here an implied interaction between two magnetic fields, rather a clever idea for a magnetician of the sixteenth century. In one case, the reaction is between the field of the earth and that of the moon, compelling the latter to rotate round its primary once every month; and the second, between the field of the earth and that of the sun, compelling our planet to revolve round the center of our system once every year.

Though an inefficient cause of the annual motion of our planet, this interaction of two magnetic fields had, nevertheless, something in common with the idea of the mutual action of material particles postulated in the Newtonian theory of universal gravitation.

This magnetic assumption by which Gilbert sought to defend the theory of the universe propounded by Copernicus was a very vulnerable point in his astronomical armor which was promptly detected and fiercely assailed by a galaxy of continental writers; all of them churchmen, physicists and astronomers of note. They accepted Gilbert's electric and magnetic discoveries and warmed up to his experimental method; they did not discard his theory of terrestrial magnetism, but rejected and scoffed at the use which he made of it to justify the heliocentric theory. They poked fun at the English philosopher for his magnetic hypothesis of planetary rotation and revolution, and succeeded in discrediting the Copernican doctrine. Error prevailed for a time, but Newton's *Principia*, published in 1687, gave the Ptolemaic system the *coup de grâce*. Gilbert's hypothesis of the interaction of planetary magnetic fields gave way to universal gravitation, and Copernicanism was finally triumphant.

Throughout the pages of Gilbert's treatise, he shows himself remarkably chary in bestowing praise, but surprisingly vigorous in denunciation. St. Thomas is an instance of the former, for it is said that he gets at the nature of the lodestone fairly well; and it is admitted that "with his godlike and perspicacious mind, he would have developed many a point had he been acquainted with magnetic experiments." Taisnier, the Belgian, is an example of the latter, whose plagiarism from Peregrinus wrings from our indignant author such withering words as "May the gods damn all such sham, pilfered, distorted works, which so muddle the minds of students!"

Besides his treatise on the magnet, Gilbert is the author of an extensive work entitled, "De Mundo Nostro Sublunari," in which he defends the modern system of the universe propounded by Copernicus and gives his views on important cosmical problems. This work was published after the author's death, first at Stettin in 1628, and again at Amsterdam in 1651.

Chancellor Bacon was well acquainted with this treatise of our philosopher; indeed he had in his collection the only two manuscript copies ever made, one in Latin and the other in English, a very singular and significant fact in view of the Chancellor's attitude toward Gilbert. Putting it crudely, one would like to know how he obtained possession of the manuscripts and what was his motive in keeping them hidden away from the philosophers of the day. "It is considered surprising," writes Prof. Silvanus P. Thompson, "that Bacon, who had the manuscripts in his possession and held them for years unpublished, should have written severe strictures upon their dead author and his methods, while at the very same time posing as the discoverer of the inductive method in science, a method which Gilberd (Gilbert) had practised for years before."

That Bacon was no admirer of Gilbert's physical and cosmical theories the following passages will show. In the "Novum Organum" the Chancellor wrote: "His philosophy is an instance of extravagant speculation founded on insufficient data"; again, "As the alchemists made a philosophy out of a few experiments of the furnace, Gilbert, our countryman, hath made a philosophy out of the lodestone" ("The Advancement of Learning"); lastly, "Gilbert hath attempted a general system on the magnet, endeavoring to build a ship out of materials not sufficient to make the rowing-pins of a boat" ("De Augmentis Scientiarum").

One is tempted to ask how this strange disregard which Bacon entertained for the scientific views of the greatest natural philosopher of his age and country came to exist? Was it due to a feeling of jealousy that could not brook a rival in the domain of the higher philosophy, or was it because Bacon, the anti-Copernican, wanted to write down Gilbert, the defender of the heliocentric theory, in the British Isles?

When reading Bacon's depreciatory remarks we have to remember that his mathematical and physical outfit was very limited even for the age in which he lived; from which it is safe to infer that he was but little qualified to pass judgment on the value of the electric and magnetic work accomplished in the workshops at Colchester or on the theories to which they gave rise.

Bacon deserves praise for denouncing the prevalent system of natural philosophy which was mainly authoritative, speculative and syllogistic instead of experimental, deductive and inductive, but he was inconsistent and forgetful of his own principles when he belittled the greatest living enemy of mere book-learning, and the most earnest advocate, by word and example, of the laboratory methods for the advancement of learning.

To avoid misapprehension, it should be here stated that Bacon was not always censorious in his treatment of his illustrious fellow-citizen, for in several places he writes approvingly of the electric and magnetic experiments contained in *De Magnete,* which he calls in his *Advancement of Learning,* "a painfull (*i.e.,* painstaking) experimentall booke." In other places he draws so freely on Gilbert without acknowledgment as to come dangerously near the suspicion of plagiarism.

Gilbert died, probably of the plague, in the sixtieth year of his age, on December 10th, 1603, and was buried in the chancel of Holy Trinity Church, Colchester, where a mural tablet records in Latin the chief facts of his life.

Dr. Fuller in his "Worthies of England" (1662) describes Gilbert as tall of stature and cheerful of "complexion," a happiness, he quaintly remarks, not ordinarily found in so hard a student and retired a person." Concluding his appreciation of the philosopher, Fuller writes: "Mahomet's tomb at Mecha is said strangely to hang up, attracted by some invisible loadstone; but the memory of this Doctor will never fall to the ground, which his incomparable book *De Magnete* will support to eternity."

Animated by a similar spirit of national pride, Dryden wrote

Gilbert shall live till loadstones cease to draw, Or British fleets the boundless ocean awe.

We shall close these remarks by Hallam's estimate of Gilbert as a scientific pioneer, contained in his *Introduction to the Literature of Europe*. "The year 1600," he says, "was the first in which England produced a remarkable work in physical science; but this was one sufficient to raise a lasting reputation for its author. Gilbert, a physician, in his Latin treatise on the magnet, not only collected all the knowledge which others had possessed on the subject, but became at once the father of experimental philosophy in this island; and, by a singular felicity and acuteness of genius, the founder of theories which have been revived after a lapse of ages and are almost universally received into the creed of science."

For well-nigh three hundred years, *De Magnete* remained untranslated, being read only by the scholarly few. The first translation was made by P. Fleury Mottelay, of New York, and published by Messrs. Wiley and Sons in the year 1893. Mr. Mottelay has given much attention to the bibliography of the twin sciences of electricity and magnetism, as the foot-notes which he has added to the translation abundantly prove.

A second translation appeared in the tercentenary year, 1900, and was the work of the members of the Gilbert Club, London, among whom were Dr. Joseph Larmor and Prof. Silvanus P. Thompson. It is a page-for-page translation with facsimile illustrations, initial letters and tail-pieces.

As one would infer from the numerous references contained in *De Magnete*, Gilbert had a considerable collection of valuable books, classical and modern, bearing on the subject of his life-work; but these, as well as his terrellas, globes, minerals and instruments, perished in the great fire of London, 1666, with the buildings of the College of Physicians, in which they were located.

A portrait of Gilbert was preserved in the Bodleian Library, Oxford, for many years; but has long since disappeared from its walls. On the occasion of the three hundredth anniversary (1903) of Gilbert's death, a fine painting representing the Doctor in the act of showing some of his electrical experiments to Queen Elizabeth and her court (including Sir Walter Raleigh, Sir Francis Drake and Cecil, Lord Burleigh, famous Secretary of State), was presented to the Mayor of Colchester by the London Institute of Electrical Engineers. A replica of the painting was sent to the St. Louis Exposition, 1904, where it formed one of the attractions of the Electricity Building.

The house in which Gilbert was born (1544) still stands in Holy Trinity Street, Colchester, where it is frequently visited by persons interested in the history of electric and magnetic science.

CHAPTER III. FRANKLIN AND SOME CONTEMPORARIES.

As already seen, the writers of Greece and Rome knew little about the lodestone; we have now to add that the knowledge of electricity which they possessed was of the same elementary character. They knew that certain resinous substances, such as amber and jet had, when rubbed, the property of attracting straws, feathers, dry leaves and other light bodies; beyond this, their philosophy did not go. The Middle Ages added little to the subject, as the Schoolmen were occupied with questions of a higher order. The Saxon Heptarchy came and went, Alcuin taught in the schools of Charlemagne, Cardinal Langton compelled a landless and worthless king to sign Magna Charta, universities were founded with Papal sanction in Italy, France, Germany, England and Scotland, Copernicus wrote his treatise on the revolution of heavenly bodies and dedicated it to Pope Paul III., Tycho Brahé made his famous astronomical observations at Uranienborg and befriended at Prague the penniless Kepler, and Columbus gave a New World to Castile and Leon—all this before the man appeared who, using amber as guide, discovered a new world of phenomena, of thought and philosophy. This man was no other that Gilbert, whose discoveries in magnetism were described in an earlier chapter. The trunk line of his work was magnetism; electricity was only a siding. One was the main subject of a life-long quest while the other was only a digression. It was a digression in which the qualities of the native-born investigator are seen at their very best: alertness and earnestness, resourcefulness and perseverance, all rewarded by a rich harvest of valuable results. It is refreshing and inspiring to read the Second Book of Gilbert's treatise, *De Magnete*, in which are recorded in quick succession the twenty important discoveries which he made in his new field of labor.

At the very outset, he found it necessary to invent a recording instrument to test the electrification produced by rubbing a great variety of substances. This he appropriately called a *versorium*; we would call it an electroscope. "Make to yourself," he says, "a rotating needle of any sort of metal three or four fingers long and pretty light and poised on a sharp point." He then briskly rubs and brings near his versorium glass, sulphur, opal, diamond, sapphire, carbuncle, rock-crystal, sealing-wax, alum, resin, etc., and finds that all these attract his suspended needle, and not only the needle, but everything else. His words are remarkable: "All things are drawn to electrics." Here is a great advance on the amber and jet, the only two bodies previously known as having the power to attract "straws, chaff and twigs," the usual test-substances of the ancients. Pursuing his investigations, he finds numerous bodies which perplex him, because when rubbed they do not affect his electroscope. Among these, he enumerates: bone, ivory, marble, flint, silver, copper, gold, iron, even the lodestone itself. The former class he called *electrica,* electrics; the latter was termed *anelectrica,* non-electrics.

To Gilbert we, therefore, are indebted for the terms electric and electrical, which he took from the Greek name for amber instead of succinic and succinical, their Latin equivalents. The noun electricity was a coinage of a later period, due probably to Sir Thomas Browne, in whose *Pseudodoxia Epidemica,* 1646, it occurs in the singular number on page 51 and in the plural on page 79. It may interest the reader to be here retold that we owe the chemical term *affinity* to Albertus Magnus, *barometer* to Boyle, *gas* to van Helmont, *magnetism* to Barlowe, magnetic *inclination* to Bond, electric *circuit* to Watson, electric *potential* to Green, *galvanometer* to Cumming, *electro-magnetism* to Kircher, *electromagnet* to Sturgeon, and *telephone* to Wheatstone.

Gilbert was perplexed by the anomalous behavior of his non-electrics. He toiled and labored hard to find out the cause. He undertook a long, abstract, philosophical discussion on the nature of bodies which, from its very subtlety, failed to reveal the cause of his perplexing anomaly. Gilbert failed to discover the distinction between conductors and insulators; and, as a consequence, never found out that similarly electrified bodies repel each other. Had he but suspended an excited stick of sealing-wax, what a promised land of electrical wonders would have unfolded itself to his vision and what a harvest of results such a reaper would have gathered in! From solids, Gilbert proceeds to examine the behavior of liquids, and finds that they, too, are susceptible of electrical influence. He notices that a piece of rubbed amber when brought near a drop of water deforms it, drawing it out into a conical shape. He even experiments with smoke, concluding that the small carbon particles are attracted by an electrified body. Some years ago, Sir Oliver Lodge, extending this observation, proposed to lay the poisonous dust floating about in the atmosphere of lead works by means of large electrostatic machines. He even hinted in his Royal Institution lecture that they might be useful in dissipating mists and fogs, and recommended that a trial be made on some of our ocean-steamers.

Gilbert next tries heat as an agent to produce electrification. He takes a red-hot coal and finds that it has no effect on his electroscope; he heats a mass of iron up to whiteness and finds that it, too, exerts no electrical effect. He tries a flame, a candle, a burning torch, and concludes that all bodies are attracted by electrics save those that are afire or flaming, or extremely rarefied. He then reverses the experiment, bringing near an excited body the flame of a lamp, and ingenuously states that the body no longer attracts the pivoted needle. He thus discovered the neutralizing effect of flames, and supplied us with the readiest means that we have to-day for discharging non-conductors.

He goes a step further; for we find him exposing some of his electrics to the action of the sun's rays in order to see whether they acquired a charge; but all his results were negative. He then concentrates the rays of the sun by means of lenses, evidently expecting some electrical effect; but finding none, concludes with a vein of pathos that the sun imparts no power, but dissipates and spoils the electric effluvium.

Professor Righi has shown that a clean metallic plate acquires a positive charge when exposed to the ultraviolet radiation from any artificial source of light, but that it does not when exposed to solar rays. The absence of electrical effects in the latter case is attributed to the absorptive action of the atmosphere on the shorter waves of the solar beam.

Of course Gilbert permits himself some speculation as to the nature of the agent with which he was dealing. He thought of it, reasoned about it, pursued it in every way; and came to the conclusion that it must be something extremely tenuous indeed, but yet substantial, ponderable, material. "As air is the effluvium of the earth," he says, "so electrified bodies have an effluvium of their own, which they emit when stimulated or excited"; and again: "It is probable that amber exhales something peculiar that attracts the bodies themselves."

These views are quite in line with the electronic theory of electricity in vogue to-day, which invests that elusive entity with an atomic structure. It is held that the tiny particles or electrons that are shot out from the cathode terminal of a vacuum tube with astounding velocity are none other than particles of negative electricity, pure and simple. They have mass and inertia, both of which properties are held to be entirely electrical, though quite analogous to the mass and inertia of ordinary, ponderable matter.

History shows that scientific theories have their periods of infancy, maturity and decay. When they have served their purpose, like the scaffolding of a building, they are removed from sight and stored away, say, in a limbo of discarded philosophy, for use of the historian of science or of the metaphysician writing on the nature of human knowledge. Such was the fate of Gilbert's "effluvium" theory of electricity, of the fluid theories of Dufay and Franklin, and the ether-strain theory of recent years. "Each physical hypothesis," says Prof. Fleming, "serves as a lamp to conduct us a certain stage in the journey. It illumines a limited portion of the path, throwing light before and behind for some distance; but it has to be discarded and exchanged at intervals because it has become exhausted and because its work is done."

It is a little surprising that the phenomenon of electrical repulsion should have escaped the attention of one so skilled in experimentation as Gilbert. Yet such was the case; and Gilbert even went so far as to deny its very existence, saying, "Electrics attract objects of every kind; they never repel." This error reminds one of

Gilbert's own saying that "Men of acute intelligence, without actual knowledge of facts, and in the absence of experiment, easily slip and err." Just twenty-nine years after Gilbert had penned this aphorism, there appeared in Ferrara an extensive work on electric and magnetic philosophy, by the Jesuit Cabeo, in which this electrical repulsion was recognized and described. Having rubbed one of his electrics, Cabeo noticed that it attracted grains of dust at first and afterward repelled them suddenly and violently. In the case of threads, hairs or filaments of any kind, he observed that they quivered a little before being flung away like sawdust. This self-repelling property of electricity, described in the year 1629, opened up a new field of inquiry, which was actively explored by a number of brilliant electricians in England and on the Continent.

This was especially the case after the building of the first frictional machine by Otto von Guericke in 1672. The burgomaster of Magdeburg had already acquired European fame by the original and sensational experiments on atmospheric pressure which he made in presence of the Emperor and his nobles in solemn diet assembled (1651). Von Guericke seems to have been of a mind with Gilbert concerning writers on natural science who treat their subjects "esoterically, miracle-mongeringly, abstrusely, reconditely, mystically"; for he affirms that "oratory, elegance of diction or skill in disputation avails nothing in the field of natural science."

Von Guericke's machine consisted of a ball of sulphur, with the hand of the operator or assistant as rubber. Some years later, the sulphur ball was replaced by Newton (some say Hauksbee) by a glass globe, which, in turn, was exchanged for a glass cylinder by Gordon, a Scotch Benedictine, who was Professor of natural philosophy in the University of Erfurt. In 1755, Martin de Planta, of Sus, in Switzerland, constructed a plate-machine which was subsequently improved by Ramsden of London. The frictional machine, as it was rightly called, has been superseded by the influence machine, a type of static generator which is at once efficient, reliable and easy of operation. The best known form for laboratory use is that of Wimshurst (1832-1903), of London.

Andrew Gordon, the Scotch Benedictine to whom reference has just been made, was a man of an inventive turn of mind. Besides, the cylindrical electric machine which he constructed, he devised several ingenious pieces of electrical apparatus, among which are the *electric chimes* usually ascribed to Franklin. They are fully described in his *Versuch einer Erklärung der Electricität*, published in 1745. On page 38, he says that he was led to try an electrical method of ringing bells; and then adds: "For this purpose I placed two small wine-glasses near each other, one of which stood on an electrified board, while the other, placed at a distance of an inch from it, was connected with the ground. Between the two, I suspended a little clapper by a silk thread, which clapper was attracted by the electrified glass and then repelled to the grounded one, giving rise to a sound as it struck each glass. As the clapper adhered somewhat to the glasses, the effect on the whole was not agreeable. I, therefore, substituted two small metallic gongs suspended one from an electrified conductor and the other from a grounded rod, the gongs being on the same level and one inch apart. When the clapper was lowered and adjusted, it moved at once to the electrified bell, from which it was driven over to the other, and kept on moving to and fro, striking the bell each time with pleasing effect until the electrified bell lost its charge." In the illustration, *a* is connected with the electrified conductor; *b* is the insulated clapper; *c* the grounded gong.

Gordon's book was published in Erfurt in 1745, while the year 1752 is that in which Franklin applied the chimes to his experimental rod to apprise him of the approach of an electric storm, an application which was original and quite in keeping with the practical turn of mind that characterized our journeyman-printer, philosopher and statesman. Unquestionably, Franklin had all the ingenuity and constructive ability needed to make such an appliance; but there is no evidence that he actually invented it. Though Franklin neither claimed nor disclaimed the chimes as his own, all his admirers would have preferred less reticence on his part when the discoveries and inventions of contemporary workers in the electrical field were concerned. He had attained sufficient eminence to permit him to look appreciatingly and encouragingly on the efforts of others.

Gordon also invented a toy electric motor in which rotation was effected by the reaction of electrified air-particles escaping from a number of sharp points. One of these motors consisted of a star of light rays cut from a sheet of tin and pivoted at the center, with the ends of the rays slightly bent aside and all in the same direction. When electrified, Gordon noticed that the star required no extraneous help to set it in motion. It was a self-starting electric-motor. In the dark, the points were tipped with light, and as they revolved traced out a luminous circle which "could neither be blown out nor decreased."

The reader will recognize in this description taken from Gordon's *Versuch*, page 45, the *electric whirl* of the lecture-table; Gordon's name is never associated with it, but that of Hamilton (Hamilton's "fly" or Hamilton's "mill") sometimes is!

This irrepressible monk seems to have been one of the earliest electrocutors, for it is said that many an innocent chaffinch fell victim to discharges from his machine; and we would be disposed to think of him as a wizard on learning that he ignited spirits by using an electrified stream of water, to the astonishment and mystification of the spectators.

Abbé Menon was kinder to the feathered tribe than his black-cowled brother of Erfurt; he did not subject them to a powerful discharge, but rather to a gentle electrification for the purpose of determining what physical or physiological effect the agent would have on the animal system. The Abbé found that cats, pigeons, sparrows and chaffinches lost weight by being electrified for five or six hours at a time, from which he concluded that electricity augments the slow, continuous perspiration of animals. The same was found to take place with the human body itself. The reader will remember that Stephen Gray in 1730 suspended a boy by means of silken cords for the purpose of electrification; Abbé Nollet did the same, and doubtless his friend Abbé Menon adopted a similar mode of insulation for complacent electrical subjects. An easier mode of operating would have been to make the child stand on a cake of resin, the insulating property of which had been discovered by Stephen Gray.

About this time, 1746, Franklin appears on the scene, and though he devoted but nine years (1746-1755) of his life to the study of electricity, he made discoveries in that fascinating branch of human knowledge that will hand his name down the centuries.

Franklin's life is interesting and instructive on account of the difficulties which he met and overcame, for his strength of will, tenacity of purpose, the philosophy which he followed, his devotedness to science, and the success which he achieved.

Our philosopher's moral code comprised the thirteen virtues of temperance, silence, order, resolution, frugality, industry, sincerity, justice, moderation, cleanliness, tranquility, chastity and humility. To each of these virtues Franklin attached a precept which makes edifying reading even at the present day: *temperance,* eat not to dullness, drink not to elation; *silence,* speak not but what may benefit others or yourself, avoid trifling conversation; *order,* let all your things have their places, let each part of your business have its time; *resolution,* resolve to perform what you ought, perform without fail what you resolve; *frugality,* make no expense, but do good to others or yourself, *i.e.,* waste nothing; *industry,* lose no time, be always employed in something useful, cut off all unnecessary actions; *sincerity,* use no hurtful deceit, think innocently and justly, and if you speak, speak accordingly; *justice,* wrong no one by doing injury or omitting the benefits that are your duty; *moderation,* avoid extremes, forbear resenting injuries so much as you think they deserve; *cleanliness,* tolerate no uncleanliness in body, clothes or habitation; *tranquility,* be not disturbed by trifles or accidents common or unavoidable; *chastity* (no remark); *humility,* imitate Jesus.

This last virtue seems to have given Franklin very much concern; for he admits that he had the appearance of humility, and immediately adds that in reality there is no passion of the human breast so hard to subdue as pride. He is shrewd enough to say that "even if I could conceive that I had completely overcome it, I should probably be proud of my humility." Like many another, the virtue which gave him the most trouble was *order*, and this never became conspicuously apparent at any time of his long life.

In his endeavors after the higher life, he seems to have been animated with the earnest spirit of the ascetic who binds himself to strive after perfection as laid down in the maxims and counsels of the Gospel. It is not without surprise and perhaps a feeling too of self-condemnation, that we read the means which he adopted to reach a high moral standard. Taking for granted that he had a true appreciation of right and wrong, he did not see why he should not always act according to the dictates of conscience. To improve himself morally and advance in the higher life, he adopted a means that should have proved effective. Taking the first of the thirteen fundamental virtues, he applied himself to its acquisition for a whole week together, after which he took the second, then the third, and so on with the rest. He thought that by making daily acts of the virtue, it would become habitual with him at the end of the week. When the last of the thirteen virtues had received its share of attention, he returned to the first one on the list and proceeded round the cycle again. Being a man of purpose and tenacity, he completed the circle of his chosen virtues four times a year; subsequently he extended the time of individual practise so as to take a whole year for the course; and later on, he devoted several years to the completion of his list.

As an aid in this work of self-betterment, Franklin examined himself daily, registering his failures in a little book which was ruled for the purpose, a column being allowed for each day and a line for each of the thirteen virtues. He naively tells us the result of this exercise of daily introspection in these words: "I am surprised to find myself so much fuller of faults than I had imagined; but I had the satisfaction of seeing them diminish."

The evening examination of conscience was always concluded by the following prayer written by Franklin himself: "O powerful Goodness! bountiful Father! merciful Guide! increase in me that wisdom which discovers my truest interest. Strengthen my resolutions to perform what that wisdom dictates. Accept my kind offices to Thy other children as the only return in my power for Thy continual favors to me."

An extensive reader, Franklin found in Thomson's poems some lines that appealed to him very strongly by the beauty of the sentiment expressed. He called them "a little prayer," which he recited from time to time:

"Father of light and life, Thou Lord Supreme, Oh, teach me what is good; teach me Thyself. Save me from folly, vanity and vice; From every low pursuit; and fill my soul With knowledge, conscious peace and virtue pure; Sacred, substantial, never-failing bliss!"

His was a praiseworthy attempt at emancipating himself from the thraldom of passion and raising himself to the high plane of perfection required by the Master when He said "Follow Me." Doubtless, as time wore on, he must have felt as many before and since, that the spirit is willing but the flesh is weak.

In his autobiography, Franklin attributes his success in business not only to his self-control, uniformity of conduct, philosophical indifference to slight or pique, but also to his habits of frugality, the result in part of his early training. "My original habits of frugality continuing," he says, "and my father having frequently repeated a proverb of Solomon, 'Seest thou a man diligent in his business? he shall stand before kings,' I from thence considered industry as a means of obtaining wealth and distinction, which encouraged me, tho' I did not think that I should ever literally *stand before kings*, which, however, has since happened." Our aged philosopher proceeds to tell us of his good fortune with a little bit of pardonable vanity, to which, by the way, he was never a great stranger, despite his philosophy, acquired virtue, and staid character. Referring to the kings of the earth, he informs us that he "*stood* before five, and even had the honor of *sitting down* with one to dinner."

An important event in Franklin's life was the founding by him of the first public library in the country in the year 1732. Though but twenty-six years of age, he seems to have been as well aware as any of the millionaire philanthropists of to-day, of the good that may be accomplished among common people by providing them with suitable reading matter. He watched with eagerness the progress of his experiment and was pleased with the success that crowned it. He observes that such libraries "tend to improve the conversation of Americans and to make common tradesmen and farmers as intelligent (well-informed?) as most gentlemen from other countries."

Peter Collinson, Fellow of the Royal Society of London, who had dealings with some Philadelphia merchants, was led to take an active interest in the library. This he did by sending over a number of books and papers relating to electricity together with an "electrical tube" with instructions for its use.

These literary and scientific contributions sent from London from time to time, excited much interest among the charter members of the Library Company, and principally that of Franklin himself. He had heard something of the new order of phenomena which was just then engaging the attention of European physicists. In the summer of 1746, while on a visit to Boston, his native place, he assisted at a lecture on electricity by a certain Dr. Spence, a Scotchman, who sought to illustrate the properties of electrified bodies by such experiments as could be made with glass tubes and suitable rubbers, the rudimentary apparatus available at the time. Franklin was impressed by what he saw and heard, even though he indulged in a little destructive criticism when he said that the experiments were "imperfectly made," because the lecturer was "not very expert." When Franklin wrote those words, he knew by repeated and painful experience the difficulty of getting satisfactory results from rubbing glass tubes or rotating glass globes, owing to the provoking attraction which plain, untreated glass has for moisture. Knowing this, he might have been less severe in his strictures on his friend, the peripatetic electrician.

It is evident, however, that the experiments which he witnessed surprised and pleased him, for, having shortly afterward received some electrical tubes together with a paper of instructions, from his London friend, Peter Collinson, he set to work for himself without delay. We may well say of him that what his right hand found to do, he did calmly, but with all his might. A twelve-month had not elapsed before he wrote: "I never was engaged in any study that so totally engrossed my attention and time as this has lately done; for, what with making experiments when I can be alone and repeating them to my friends and acquaintance who, from the novelty of the thing, come continually in crowds to see them, I have had little leisure for anything else." (1747.)

Here we see the calm, persistent character of the philosopher united with the affability and communicativeness of the gentleman.

For the sake of encouraging others as well, perhaps, as through a sense of personal relief, Franklin had a number of long tubes of large bore blown at the local glass-house, which tubes he distributed to his friends that they, too, might engage in research work. In this way, rubbing and rubbing of an energetic kind became quite an occupation in the Franklin circle. Kinnersley, whose name still survives in works on static electricity in connection with an electric "thermometer" which he devised, was among the band of ardent workers who ungrudgingly acknowledged Franklin's superior acumen, comprehensive grasp of detail and wondrous insight into the mechanism of the new phenomena. If we say that Franklin was not a genius, it is only for the purpose of adding that even in those early electrical studies he displayed an uncommon amount of the unlimited capacity for taking pains which is said to be associated with that brilliant gift. He tested all his results with great care and in a variety of ways before accepting any of them as final; and considered his explanations of them provisional, being ever ready to modify them or give them up altogether if shown to conflict with the simple workings of nature.

As early as 1733, the refined and tactful Dufay, in France, showed by numerous experiments on woods, stones, books, oranges and metals that all solid bodies were susceptible of electrification. This was a notable advance which swept away Gilbert's classification of bodies into electrics and non-electrics. The French physicist soon drew from his observations the conclusion that electrification produced by friction is of two kinds, to which he applied the terms vitreous and resinous, the former being developed when glass is rubbed with silk and the latter when amber or common sealing-wax is rubbed with flannel. He noticed, too, that silk strings repelled each other when both were touched either with excited glass or sealing-wax; but that they attracted each other when touched one with glass and the other with sealing-wax. From these observations, he deduced the electrostatic laws, that similarly electrified bodies attract while dissimilarly electrified bodies repel each other.

The law of distance was discovered later by Coulomb, who, in 1785, showed that the law of repulsion as well as of attraction between two electrified particles varies inversely as the square of the distance. In the year 1750, the law of the inverse square for magnets was stated by John Michell, who expressed it by saying that the "attraction and repulsion decrease as the square of the distance from the respective poles increases." Michell was fourth wrangler of his year (1748-9), Fellow of Queen's College, Cambridge, and inventor of the *torsion balance*, which, however, he did not live to use; but which, in the hands of Cavendish, yielded important results on the mean density of the earth. Coulomb probably re-invented the "balance" and applied the practical, laboratory instrument which he made it, to the study of the quantitative laws of electricity and magnetism.

To observe and correlate phenomena is the special work of the physicist; to speculate on ultimate causes is the privilege of the philosopher. Dufay was both. The theory which he offered was a simple one, even if untrue to nature. It was a good working hypothesis for the time being.

According to this theory, there are two distinct, independent electrical fluids mutually attractive but self-repelling. With that postulate, Dufay was able to offer a plausible explanation of a great many phenomena that puzzled the electricians of the time.

Franklin, however, held a different view; rejecting the dual nature of electricity, he propounded his one-fluid theory, which was found equally capable of explaining electrical phenomena. A body having an excess of the fluid was said to be *positively* charged, while one with a deficit was said to be *negatively* charged. The sign plus was used in one case and the sign minus in the other; and just as two algebraical quantities of equal magnitude but opposite sign give zero when added together, so a conductor to which equal quantities of positive and negative electricity would be given would be in the neutral state. The Franklinian theory was welcomed in England, Germany and Italy, but it met with opposition in France from the brilliant Abbé Nollet and the followers of Dufay.

Each of the rival theories affords a mental conception of the forces in play and also a consistent explanation of the resulting phenomena. Their simplicity, and, at the same time, the comprehensiveness of explanation which they afford, will continue to give them a place in our text-books for many years to come.

Efforts are being made to apply the *electronic* theory to the various phenomena of electrostatics, the electron being the smallest particle of electricity that can have separate, individual existence. It is many times smaller than the hydrogen atom, the smallest of chemical atoms, and it possesses all the properties of negative electricity. By the loss of one or more electrons, a body becomes positively electrified, whereas by the acquisition of one or more electrons it becomes negatively electrified. The electron at rest gives rise to the phenomena of electrostatics; in motion, it gives rise to electrical currents, electromagnetism and electric radiation.

We do not know what led Franklin to call positive the electrification of glass when rubbed with silk, and negative that of sealing-wax when rubbed with flannel. If he meant to imply that positive is the more important of the two, he erred, for many reasons can be given to show the preponderating influence of negative electricity; but it is too late now to change the terminology.

If asked to point out differences between the physical effects of positive and negative electrification, we would refer to the positive brush, which is finer and much more developed than the negative; to the Wimshurst machine, with its positive brushes on one side and negative "beads" on the other; to the positive charge acquired by a clean plate of zinc when exposed to ultraviolet light; to the ordinary vacuum tube in which there is a violet glow at the cathode end or negative terminal; to Crookes's tubes, X-ray tubes and other high vacuum tubes, in which electrified particles, Kelvin's *molecular torrent*, are shot out from the negative electrode with great velocity; and to arc-lamps using a direct current in which the plus carbon is hollowed out crater-like, has the higher temperature and wastes away twice as fast as the negative.

The year 1746 is an *annus mirabilis* in the history of electricity, for it was in the January of that year that an attempt to electrify water by Musschenbroek, of Leyden, led to the discovery of the principle of the electrostatic condenser. Whatever may be thought of the claim for priority put forward in favor of Dean von Kleist, of Cammin in Pomerania, or of Cunæus, of Leyden, it is certain that the discovery became known throughout Europe by the startling announcement and sensational description given of it by Musschenbroek, a renowned professor of a renowned university. He was not only surprised but terror-stricken by the effect of the electric energy which he had unconsciously stored up in his little phial; for after telling his French friend Réaumur, the physicist, that he felt the commotion in his arms, shoulders and chest, he added that he would not take another shock for the whole kingdom of France! A resolution destined to be broken, like so many others before and since.

Very different was the sentiment of Bose, Professor of Physics in the University of Wittenberg, who is credited with saying that he would like to die by the electric shock, that he might live in the memoirs of the French Academy of Sciences.

The Leyden jar became at once the scientific curiosity and universal topic of discussion of the time; and not only was it the curiosity, but also the *crux* of the day, puzzling investigators, perplexing philosophers and giving rise to animated controversies. The mystery was soon dispelled, however, when Franklin began in 1747 his searching inquiry into the electric conditions of each element of the jar. Nothing escaped his subtle mind and nothing was left undone by his deft hand. The evidence of experiment and the logic of facts carried at last conviction even with Londoners and Parisians, who were wont to look upon Americans as mere colonists, who had neither time nor opportunity for scientific pursuits, being obliged to hew their way through virgin forests or drive the roving Indian back from their frontiers into the wilds of the West. The theory of the Leyden jar given by Franklin 160 years ago has stood the test of time. It has met with universal acceptance; and, despite our manifold advances, but little of permanent value has been added to it.

It is very interesting to follow the main lines of this magnificent research. Franklin electrifies, in the usual way, water contained in a small flask, complaisantly taking the shock on completing the circuit. To find where the charge resides, whether in the hand of the operator, as some said, or in the water, as others maintained, he again electrifies the water and pours it into another flask, which fails, however, to give a shock, thus showing that the charge had not been carried over with the water. Convinced that the charge was still somewhere in the first phial, he carefully poured water into it again; and found, to his intense satisfaction, that it was capable of giving an excellent shock. It was now clear to him that the energy of the charge was either in the hand of the experimenter or in the glass itself, or in both. To determine this nice point, he proceeds to construct a "jar" which could easily be taken to pieces. For this purpose, he selected a pane of glass; and, laying it on the extended hand, placed a sheet of lead on its upper surface. The leaden plate was then electrified; and when touched with the finger, a spark was seen and a shock felt. By the addition of another plate to the lower surface, the shocking power of this simple condenser was increased. In this efficient form he had a readily dissectible condenser, which allowed him to throw off and replace the coatings at will, and thereby to prove beyond cavil that the seat of the stored-up electric energy is not in the conductors, but in the glass itself. This was a discovery of the first magnitude and one destined to associate the name of Franklin with those of the most eminent electricians down the ages. Fig. 11 shows the modern form of the jar with movable coatings.

In the "fulminating" pane, as it came to be called, we have one of the eleven elements of Franklin's historic battery of 1748. It is interesting to notice that he was accustomed to connect his "panes" in series while charging (Fig. 12), but that he preferred to join similar coatings together, that is, to couple them in "parallel" (Fig. 13), for powerful discharges. Fig. 14 shows three jars in "parallel."

Later on, he arranged Leyden jars so that the inside coating of one could be hooked to the outside coating of another, the first of the series hanging down from the prime conductor of the machine, while the last one was grounded. "What is driven out of the tail of the first," he quaintly says, "serves to charge the second; what is driven out of the second serves to charge the third, and so on." This has become known as the "cascade" method of charging a battery, owing to the flow of electricity from one jar to the next (Fig. 15). Electricians, however, have discarded the picturesque "cascade" for the prosaic term of "series" or "tandem" arrangement.

Franklin also noticed that a phial cannot be charged while standing on wax or on glass, or even while hanging from the prime conductor, unless communication be formed between its outer coating and the floor, the reason given being that "the jar will not suffer a charging unless as much fire can go out of it one way as is thrown in by the other." (1748.)

Following his very ingenious Philadelphia friend and co-worker, Kinnersley, he varies the mode of charging by electrifying the outside of the jar and grounding the inner coating; for "the phial will be electrified as strongly if held by the hook and the coating applied to the globe as when held by the coating and the hook applied to the globe." (1748.)

The globe here referred to is the glass globe of Franklin's frictional machine of American make, which, when rotated, was electrified positively by contact with the hand or with a leather rubber. Franklin also used a sulphur ball or "brimstone" globe, and observed that the electrification produced on it differed in kind from that developed on the glass globe. (1752.)

It may here be stated that the first to use a *leather cushion* as a substitute for the hand in the frictional machine, was Winkler, of Leipzig (1745); the efficiency of the rubber was increased by Canton, of London, who covered it with an *amalgam* of tin and mercury (1762). Bose, of Wittenberg, had previously added the *prime-conductor*, which greatly augmented the electrical capacity and output of the machine.

In 1750 Franklin imitated the effect of lightning on the compasses of a ship by the action of a jar discharge on an unmagnetized steel needle. "By electricity," he says, "we have frequently given polarity to needles and reversed it at pleasure."

Similar experiments are made to-day in every lecture-course on static electricity; but the experimenter, when wise, does not announce beforehand which end of the needle will be north and which south, as he is just as likely to be wrong as right, the uncertainty being due to the fact that the discharge of a Leyden jar is not a current of electricity in one direction, but rather a few sudden rushes or rapid surgings of electricity to and fro; in other words, it is oscillatory in character instead of being continuous in one direction.

Franklin did not know this; although he made a very pertinent remark in 1749 when he likened the mechanical condition of the glass of a charged jar to that of a bent rod or a stretched spring. "So, a straight spring," he says, "when forcibly bent must, to restore itself, contract that side which in the bending was extended, and extend that side which was contracted." Franklin knew, of course, that the bent rod, when released, would swing to and fro a few times before settling down to its state of rest; but he failed to see the analogy between it and the strained glass of the charged Leyden jar.

It is to Joseph Henry (1799-1878), the Faraday of America, that we owe the recognition and statement of the oscillatory character of the discharge from Leyden jars and condensers generally. He discovered and published this cardinal fact in 1842. His words deserve recording. "The discharge, whatever may be its nature, is not correctly represented (employing for simplicity the theory of Franklin) by the single transfer of an imponderable fluid from one side of the jar to the other; the phenomenon requires us to admit *the existence of a principal discharge in one direction and then several reflex actions backward and forward, each more feeble than the preceding, until equilibrium is attained.*" The italics are Prof. Henry's.

It is precisely this oscillatory character of the spark-discharge that enables us to send out trains of electric waves into the all-pervading ether, and thus to communicate, by "wireless," with remote stations.

Having conclusively proved that the energy of a charged condenser resides in the dielectric, Franklin next tries to find whether "the electric matter" in the case of conductors is limited to the surface or whether it penetrates to an appreciable depth. To ascertain this, he insulates a silver fruit-can and brings a charged ball, held by a silk thread, into contact with the outer surface. On testing after removal, he found that the ball retained some of its charge, whilst it lost all if allowed to touch the bottom of the vessel. Surprised at this unexpected difference, he repeated the experiment again and again, only to find the ball every time without a trace of charge after contact with the interior of the vessel. This perplexed and puzzled him. "The fact is singular," he says, "and you require the reason? I do not know it. I find a frank acknowledgment of one's ignorance is not only the easiest way to get rid of a difficulty, but the likeliest way to obtain information, and therefore I practice it. I think it an honest policy. Those who affect to be thought to know everything, often remain long ignorant of many things that others could and would instruct them in, if they appeared less conceited."

This was in 1755. Cavendish in 1773 and Coulomb in 1788 independently attacked the same problem; and having proved by their classic experiments that a static charge is limited to the surface of conductors, it was but a step to infer that such a distribution of electricity implies that the law of force between two elements of charge, or between two point-charges, is the law of the inverse square of the distance.

It will also be remembered that Faraday, not knowing what had been accomplished eighty years before in Philadelphia, used for one of his best-known experiments an ice-pail, into which he lowered an electrified ball for the purpose of showing the exact equality of the induced and the inducing charge. The similarity of apparatus and mode of procedure are remarkable.

In pursuing his work, Franklin placed a charged jar on a cake of wax and other insulating materials, and drew sparks from it by touching successively the knob and the outer coating, repeating the process a great number of times to his infinite delight. He next attached a brass rod to the outside, bending it and bringing the other end close to the knob (Fig. 16) connected with the inner coating. Between these two he suspended a leaden ball by a silk thread and found, as he expected, that it played to and fro between the terminals for a considerable time. Observe that we have here a definite mass maintained in a state of reciprocating motion by a series of electric attractions and repulsions. We have in fact an electro-motor, closely resembling the star and the chimes of Gordon, the Benedictine, 1745; a mere toy, if you will, but still a remarkable invention. We repeat the same experiment to-day only with a little more harmony, by substituting for the knobs two little bells, which emit a soft, musical note when struck by the interhanging clapper.

This experiment has further significance, for, like Gordon's chimes, it is an instance of the conveyance of electricity from one point of space to another by means of a material carrier, a mode of transfer which has since been called "electric convection," the full meaning of which was not revealed until Rowland (1848-1901), made his famous experiment of 1876 in the laboratory of the University of Berlin with a highly-charged, rapidly-revolving, ebonite disc. It was apropos of this experiment that the illustrious Clerk Maxwell, of the University of Cambridge, wrote to his friend, Professor Tait, of Edinburgh, saying that:

"The mounted disc of ebonite Had whirled before, but whirled in vain; Rowland of Troy, that doughty knight, Convection currents did obtain, In such a disc, of power to wheedle From its loved north, the needle."

We may here say that Franklin was no stranger to the work done by the electrical pioneers of the Old World, his diligent London friend, Peter Collinson, keeping him advised by means of letters, books and pamphlets, in which inspiration and practical hints must have been found. He certainly was well acquainted with the achievements of Dr. Watson and Dr. Bevis, of London, as well as with the theories and experiments of Dufay and Abbé Nollet in Paris. It is germane to the subject to say that Dr. Bevis used mercury and iron filings for the inner coating of his jars, as well as sheet lead for both. He also experimented with coated panes of glass instead of jars. About this, Franklin wrote to Collinson: "I perceive by the ingenious Mr. Watson's last book, lately received, that Dr. Bevis had used, before we had, panes of glass to give a shock; though till that book came to hand, I thought to have communicated it to you as a novelty." (1748.)

Franklin gave way to a little pleasant humor when, in 1748, he proposed to wind up the "electrical season" by a banquet à la Lucullus, to be given to a few of his friends and fellow-workers, not in a sumptuously decorated hall, but *al fresco*, on the banks of the Schuylkill. "A turkey is to be killed for our dinner by the electrical shock," he wrote, "and roasted by the electrical jack before a fire kindled by the electrical bottle, when the healths of all the famous electricians in England, Holland, France and Germany are to be drunk in electrified bumpers under the discharge of guns fired from the electrical battery."

It is hardly to be supposed that such an elaborate program was carried out. Indeed the difficulty of preparing the apparatus and getting it ready for action on the banks of a river were formidable enough to say the least. Franklin, however, had a Leyden battery capable of doing considerable electrocution, for with two jars of six gallons capacity each, he knocked six men to the ground; the same two jars sufficed to kill a hen outright, whereas it required five, he tells us, to kill a turkey weighing ten pounds.

The "electrical bumper" was a wine-glass containing an allowance, let us say, of some favorite brand and charged in the usual way. On approaching the lips the two coatings would be brought within striking-distance and a spark would take place, if not to the delight of the performer, at least to the amusement of the on-lookers. It was subsequently remarked that guests whose upper lip was adorned with a moustache could quaff the nectar with impunity, as every bristle would play the part of a filiform lightning-rod and prevent the apprehended, disruptive discharge!

Not quite so humorous was his suggestion of a hammock to be used by timid people during an electric storm: "A hammock or swinging-bed, suspended by silk cords equally distant from the walls on every side, and from the ceiling and floor above and below, affords the safest situation a person can have in any room whatever; and which, indeed, may be deemed quite free from danger of any stroke of lightning." (1767.)

In his experiments on puncturing bodies by the spark-discharge, Franklin does not fail to notice the double burr produced when paper is used. His words are:

"When a hole is struck through pasteboard by the electrified jar, if the surfaces of the pasteboard are not confined or compressed, there will be a bur raised all round the hole on both sides the pasteboard, for the bur round the outside of the hole is the effect of the explosion every way from the centre of the stream and not an effect of direction." (1753.) The spelling is Franklin's *unreformed*.

The to-and-fro nature of the discharge was thought, at a time, to account satisfactorily for the burr raised on each side of the pasteboard; but Trowbridge, of Harvard, has shown that even a unidirectional discharge, such as can be obtained by inserting a wet string or any high resistance in the circuit, would produce a double burr, from which we infer, confirming Franklin, that this effect of the discharge is caused by the sudden expansion of air within the paper itself.

By the year 1749, Franklin had reached the conclusion that the lightning of the skies is identical with that of our laboratories, basing his belief on the following analogies which he enumerates in the notes or "minutes" which he kept of his experiments: "The electric fluid agrees with lightning in these particulars: (1) Giving light; (2) color of the light; (3) crooked direction; (4) swift motion; (5) being conducted by metals; (6) crack or noise in exploding; (7) rending bodies it passes through; (8) destroying animals; (9) melting metals; (10) firing inflammable substances; and (11) sulphurous smell."

But although he felt the full force of the analogical argument, Franklin knew that the matter could not be finally settled without an appeal to experiment; and accordingly he adds: "The electric fluid is attracted by points; we do not know whether this property is in lightning. But since they agree in all the particulars wherein we can already compare them, is it not probable that they agree likewise in this? Let the experiment be made." (1749.)

In writing to Collinson in July, 1750, he tells his London friend how the experiment may be made: "On the top of some high tower or steeple, place a kind of sentry-box—big enough to contain a man—and an electrical stand. From the middle of the stand let an iron rod rise and pass, bending out of the door, and then upright 20 or 30 feet, pointed very sharp at the end. If the electrical stand be kept clean and dry, a man standing on it, when such clouds are passing low, might be electrified and afford sparks, the rod drawing fire to him from the cloud."

Collinson brought some of Franklin's letters to the notice of fellow-members of the Royal Society with a view to their insertion in the *Philosophical Transactions* of that learned body; but even his epoch-making letter to Dr. Mitchell, of London, on the identity of lightning and electricity, was dismissed with derisive laughter. The Royal Society made amends in due time for their contemptuous treatment of the American philosopher by electing him member of the Society and by awarding him the Copley medal in 1753.

Disappointed as he was, Collinson collected Franklin's letters and published them under the title of *New Experiments and Observations on Electricity made at Philadelphia in America*. The pamphlet appeared in 1751, and was immediately translated into French by M. d'Alibard at the request of the great naturalist Count de Buffon.

The experiments described in the pamphlet, and especially that of the pointed conductor, were taken up in Paris with great enthusiasm by de Buffon himself, by d'Alibard, a botanist of distinction, and by de Lor, a professor of physics. Following out the instructions given by Franklin, they were all able to report success: d'Alibard on May 10th, de Lor on May 18th, and de Buffon on May 19th, 1752.

De Buffon erected his rod on the tower of his château at Montbar; de Lor, over his house in Paris, and d'Alibard, at his country seat at Marly, a little town eighteen miles from Paris. D'Alibard was not at home on the eventful afternoon of May 10th; but before leaving Marly, he had drilled a certain Coiffier in what he should do in case an electric storm came on during his absence. Though a hardy and resolute old soldier and proud of the confidence placed in him, Coiffier grew alarmed at the long and noisy discharges which he drew from the *insulated* rod on the afternoon of May 10th. While the storm was still at its height he sent for the Prior of the place, Raulet by name, who hastened to the spot, followed by many of his parishioners. After witnessing a number of brilliant and stunning discharges, the priest drew up an account of the incident and sent it, at once, by Coiffier himself to d'Alibard, who was then in Paris. Without delay d'Alibard prepared a memoir on the subject which he communicated to the Académie des Sciences three days later, viz.: on May 13th. In the concluding paragraph, the polished academician pays a graceful tribute to the philosopher of the Western World:

"It follows from all the experiments and observations contained in the present paper, and more especially from the recent experiment at Marly-la-ville, that the matter of lightning is, beyond doubt, the same as that of electricity; it has become a reality, and I believe that the more we realize what he (Franklin) has published on electricity, the more will we acknowledge the great debt which physical science owes him."

We may, in passing, correct the error of those who credit French physicists with having originated the idea of the pointed conductor. Such writers should read the words of d'Alibard in the beginning of his memoir, where he says: "En suivant la route que M. Franklin nous a tracée, j'ai obtenu une satisfaction complète"; that is, "In following the way traced out by Franklin, I have met with complete success." To Franklin, therefore, belongs the idea of the pointed rod of 1750, which became the lightning conductor of subsequent years; to the Parisian savants belongs the great distinction of having been the first to make the experiment and verify the Franklinian view of the identity of the lightning of our skies with the electricity of our laboratories.

Franklin had precise ideas on the action of his pointed conductors, clearly recognizing their twofold function: (1) that of preventing a dangerous rise of potential by disarming the cloud; and (2) that of conveying the discharge to earth, if struck. In some of his letters, he complains of people who concentrate their attention on the preventive function, forgetting the other entirely. "Wherever my opinion is examined in Europe," he wrote in 1755, "nothing is considered but the probability of these rods preventing a stroke, which is only a part of the use I proposed for them; and the other part, their conducting a stroke which they may happen not to prevent, seems to be totally forgotten, though of equal importance and advantage."

A favorite illustration of Franklin's showing the discharging power of points, consisted in insulating a cannon ball against which rested a pellet of cork, hung by a silk thread. On electrifying the ball, the cork flies off and remains suspended at a distance, falling back at once, as soon as a needle is brought near the ball. (1747.)

He also used tassels consisting of fifteen or twenty long threads (Fig. 17), and even cotton-fleece, the filaments of which stand out when electrified, but come together when a pointed rod is held underneath. He also noticed that the filaments do not collapse when the point of the rod is covered with a small ball. (1762.)

Franklin's views on lightning-rods met with some opposition in France from the brilliant Abbé Nollet, and in England from Dr. Benjamin Wilson. The latter was mainly instrumental in bringing about the famous controversy of "Points *vs.* Knobs." In 1772, a committee was appointed by the Royal Society to consider the best means of protecting the powder-magazines at Purfleet from lightning. On the committee with Dr. Wilson were Henry Cavendish, the distinguished chemist and physicist, and Sir John Pringle, President of the Royal Society. The report favored sharp conductors against blunt ones advocated by Dr. Wilson. Five years later, in 1777, the question was again brought up, and again the new committee decided in favor of pointed terminals, convinced "that the experiments and reasons made and alleged to the contrary by Mr. Wilson were inconclusive."

Dr. Wilson, being a man of influence, succeeded in having his views taken up by the Board of Ordnance. It has been remarked that this controversy would never have attracted attention but for the fact that the discoverer of the effect of points was Franklin. He was an American and the dispute with the colonies was then at its height. The war of the Revolution had begun, and the British forces had already met with serious reverses. No patriot could, therefore, admit any good in points. George III. took sides, decreed that the points on the royal conductors at Kew should be covered with balls, and ordered Sir John Pringle to support Dr. Wilson. Sir John gave the dignified answer: "Sire, I cannot reverse the laws and operations of nature"; to which the King, incensed that so incompetent a man should hold such an important office, replied: "Then, Sir John, perhaps you had better resign," which Sir John did.

A wit of the time put the matter epigrammatically when he wrote:

"While you, great George, for knowledge hunt And sharp conductors change to blunt, The nation's out of joint; Franklin a wiser course pursues, And all your thunder useless views By keeping to the point."

It was in connection with this heated controversy that Franklin wrote the following admirable words:

"I have never entered into any controversy in defence of my philosophical opinions. I leave them to take their chance in the world. If they are *right*, truth and experience will support them; if *wrong*, they ought to be refuted and rejected. The King's changing his *pointed* conductors for *blunt* ones is, therefore, a matter of small importance to me."

It was not until September, 1752, that Franklin raised a rod over his own house. This experimental conductor was made of iron fitted with a sharp steel point and rising seven or eight feet above the roof, the other end being buried five feet in the ground. In order to avoid useless personal displacement, Franklin, the economist of time, made an automatic annunciator similar to that devised by Gordon in 1745, and described by Watson in his *Sequel*, 1746, to apprize him of the advent of a good thunder-gust. Instead of making the rod of one continuous length, it was divided on the staircase, opposite his chamber door, the ends being drawn apart to a horizontal distance of a few inches. Screwing a pair of tiny gongs to the ends, he suspended between them a brass ball, held by a silk thread, to act as clapper. Whenever a thundercloud came hovering by, the bells began to ring, thereby summoning the philosopher to his "laboratory" on the staircase.

Franklin's rod, erected over his house in the summer of 1752, was evidently intended by him for experimental rather than protective purposes. There is no doubt whatever in his mind about the use of such pointed conductors for the protection of buildings and ships against the destructive effects of lightning. He expressly says, in an article printed in *Poor Richard's Almanack* for 1753, that "It has pleased God in His infinite goodness to mankind, to discover to them the means of securing their habitations and other buildings from mischief by thunder and lightning. The method is this: provide a small iron rod (it may be made of the rod-iron used by the nailers), but of such a length, that one end being 3 ft. or 4 ft. in the moist ground, the other may be 6 ft. or 8 ft. above the highest part of the building. To the upper end of the rod fasten about a foot of brass-wire, the size of a common knitting needle, sharpened to a fine point; the rod may be secured to the house by a few small staples. If the house or barn be long, there may be a rod and point at each end, and a middling wire along the ridge from one to the other. A house thus furnished will not be damaged by lightning, it being attracted by the points and passing through the metal into the ground without hurting anything. Vessels also, having a sharp-pointed rod fixed on the top of their masts, with a wire from the foot of the rod reaching down round one of the shrouds to the water, will not be hurt by lightning."

It is well known, as Dr. Rotch, Director of the Blue Hill Observatory, recently pointed out, that the matter for these almanacs was prepared by Franklin himself under the pen-name of Richard Saunders. As the above passage appeared in the almanac for 1753, it is obvious that it must have been ready sometime toward the end of 1752. Furthermore, we know that it was actually in the hands of the printer in the middle of October of that year, for the *Pennsylvania Gazette* of Oct. 19th says that the almanac was then in press and that it would be on sale shortly. Whence it follows that the year 1752 is the year of the invention of the lightning rod, and not 1753 or 1754 as often stated.

The instructions given by Franklin include all the essentials necessary for the erection of a lightning conductor. It may be made of iron or copper, flat or round, but must make good "sky" and good "earth." The former condition is secured by screwing to the top of the rod either copper or platinum terminals ending in sharp points; and the latter, by burying the lower end deep in moist soil. Between "sky" and "earth" the rod must be continuous.

The function of the rod is twofold, as Franklin well recognized, preventive and preservative. It prevents the stroke, under ordinary conditions, by the action of the points, which send off copious streams of air and dust particles electrified oppositely to that of the cloud. Even at a distance, the dangerous potential of the cloud is reduced by these convection currents and the stroke ordinarily averted. It is clear that ten points are more efficacious than one, and fifty more than five. Hence the number of points which we see distributed over the higher and more conspicuous parts of a building, all of which are carefully connected with the lightning conductor.

However well a building may theoretically be protected, conditions will occasionally arise when the rod will inevitably be struck; its preservative function then comes into play, by which it carries the energy of the disruptive discharge safely to earth.

The experience of more than a century shows that the lightning-rod affords protection in the great majority of cases; but it would be at least a mild exaggeration to say that it never failed, even when properly constructed.

At first, the erection of lightning-rods was opposed in the New World as well as in the Old: some based their opposition to the novelty on religious grounds, saying that, as lightning and thunder are tokens of divine wrath, it would be impious to interfere in any way with their manifestations. This objection was met by saying that for a parity of reason we should avoid protecting ourselves against the inclemencies of the weather.

Others opposed the use of the rods on the score that they invited or attracted the flash, which was answered by saying that they attract lightning as much as a rain-pipe attracts a shower, and no more.

The death of Professor Richmann, of the University of St. Petersburg, also tended to retard the adoption of the rod for the protection of buildings; but the invalidity of that objection became apparent when the circumstances of the accident became known. Richmann's conductor was like d'Alibard's (1751), an experimental rod, and as such was insulated at the lower end. It was, therefore, not a lightning-rod at all, inasmuch as it was not grounded. On August 6th, 1753, during a violent electric storm, Richmann happened to be close to his exploring rod observing the indications of a roughly-made electrometer, when a sharp thunder-clap was heard, and at the same instant a ball of fire was seen by Richmann's assistant to dart from the apparatus and strike the head of the unfortunate Professor, who fell over on a near-by chest and expired instantly. His assistant was stunned for a while. On regaining consciousness, he ran to the aid of the Professor; but it was too late, the body was lifeless.

In recording this tragic event, Priestley, the historian of electricity, says that, "It is not given to every electrician to die in so glorious a manner as the justly envied Richmann."

For one, we do not "envy" Professor Richmann's fate, and we think that the phrase "tragic manner" would better suit the circumstances of his death than the "glorious manner" of Dr. Priestley.

Risks of a similar character were taken by Franklin in Philadelphia, de Romas in Bordeaux, and d'Alibard's representative at Marly, when experimenting with kites and insulated rods; they took their lives in their hands, though they may not have thought so.

A few years ago, Sir William Preece said that a man might with impunity "clasp a copper rod an inch in diameter, the bottom of which is well connected with moist earth, while the top of it receives a violent flash of lightning; the conductor might even be surrounded by gunpowder in the heaviest storm without risk or danger."

It is not on record that the English electrician ever clasped a lightning conductor or even stood in close proximity to one during an electric storm. The above statement was as sensational as it was unwise and foolhardy. The neighborhood of a rod during a storm is a zone of danger, owing to the electrical surgings which are set up in it, and, as such, is to be avoided.

The death of Richmann caused quite a sensation throughout Europe, and naturally the lightning-rod came in for severe condemnation. Among the memoirs to which the fatality gave rise was one written in the heart of Moravia and addressed to the celebrated Euler, Director of the Academy of Sciences at Berlin. The writer was a monk of the Premonstratensian Order, whose field of labor was at Prenditz.

In the year 1754, this country priest made experiments with lightning conductors on a scale that transcended anything done in Paris, London or Philadelphia. The accompanying illustrations show the conductor which Divisch (also Diwisch) raised at Prenditz (also Brenditz) in the summer of that year to demonstrate publicly the efficacy of such apparatus in breaking up thunder-clouds and neutralizing the destructive energy pent up in their electric charges. Prenditz, it would appear, suffered severely from electric storms; and it was mainly for the safety of the locality that the good priest devoted himself with earnestness to the study of electrical phenomena.

As such a man deserves to live in the memory of posterity, we have sought out the leading facts of his career mainly from Father Alphons Zák, of Pernegg, in Lower Austria, a distinguished writer of the Order to which Divisch belonged, and have woven such details as we obtained from him and others into the simple narrative that follows.

Procopius Divisch (Prokop Diwisch) was born on Aug. 1st, 1696, at Helkowitz-Senftenberg in Bohemia. He spent his youth at Znaim, where he studied the humanities and philosophy at the College conducted by the Jesuit fathers in that Moravian city. In 1719, when in his twenty-third year, he decided to quit the common ways of the world in order to lead the higher life in the Premonstratensian Order at Kloster-Bruck. At the ripe age of 30, Divisch was ordained priest, in 1726, after which he taught philosophy and theology to classes of young aspirants to the ecclesiastical state. In 1733 he went to the

University of Salzburg and won his double Doctorate in theology and philosophy. Three years later, in 1736, he was appointed parish priest of Prenditz, a small Moravian town on the road to Austerlitz, since of Napoleonic fame. Here he remained for five years, returning in 1741 to Bruck as Prior of the Kloster or monastery situated there. At the end of the Seven Years' War of the Austrian succession, he quitted Bruck, in 1745, for his parish at Prenditz, where he spent the last twenty years of his life in the pastoral ministrations of his sacred office and in electrical experimentation, of which he was very fond.

The curative property of the new agent was heralded throughout Europe about this time in terms of unmeasured praise. Some of Divisch's ailing parishioners, believing him to be an expert in electrical manipulation, applied to him for a little alleviation of their woes. The good-hearted priest did not turn them away, but thought it desirable to treat them to the therapeutic effect of such sparks as he could get from his homemade frictional machine. The results were various, depending probably on the confidence and imagination of the patient. Several remarkable cures seem to have been effected either by the electric spark or by the persuasive powers of the operator, or by both combined, with the result that people far and wide were divided in their opinion of the Pastor of Prenditz. Some physicians said that he was interfering with their practice, and even clergymen found fault with him for indulging in work which they thought unsuited to the cloth. A general impression, too, seems to have prevailed that his electrical experiments, especially those with his lightning conductor, were likely to prove harmful in more ways than one.

On the other hand, Divisch had admirers in high places, among whom were the Emperor Francis I. of Germany and his imperial consort, Maria Theresa. Having been invited to Vienna, Divisch repaired to the Austrian capital, where, with the aid of Father Franz, another electrical devotee, he gave a demonstration of the wonderful capability of the new form of energy before the grandees of the empire.

When he came to the electrical property of points, he showed their discharging power in a very original way, one which must have made his assistant uneasy for a while. At times, the machine worked by Father Franz gave excellent results; at others, it failed to generate. It was noticed by the critical few that when the machine failed, Divisch was close by; while when it worked normally, he was at some distance away. After a number of such alternations of success and failure which sorely perplexed the assistant, himself a man of renown in Vienna, Divisch explained the occurrence by saying, with a merry twinkle in his eye, that the failure of the machine to generate when he was close to it, apparently seeking out the cause of the breakdown, was due to a number of pin-like conductors which he had concealed for the purpose in his peruke and which neutralized the charge on the rotating generator!

The identity of the lightning of our skies with the artificial electricity of our laboratories was suspected by many before the middle of the eighteenth century. Englishmen like Hauksbee, Hall, Gray, Freke, Martin and Watson; Germans like Bose and Winkler, and Frenchmen like Abbé Nollet, had already published their suspicions and conjectures anent the matter. Franklin, too, had indicated twelve points of analogy between the two, in 1749, in his letter to Collinson, of London. Though he felt the force of the analogical agreement, he also felt that the matter could not be definitely settled without an appeal to experiment. Accordingly, he added: "The electric fluid is attracted by points; we do not know whether this property is in lightning. But since they agree in all the particulars wherein we can already compare them, is it not probable that they agree likewise in this? Let the experiment be made."

The experiment was made by Franklin himself by means of his kite two years later, in the summer of 1752, and also by the lightning-rod which he erected over his own house in the autumn of the same year. Doubtless Divisch heard of the marvelous effects obtained from d'Alibard's insulated conductor at Marly; at any rate, he erected in an open space at some little distance from his rectory at Prenditz, a lightning conductor 130 feet in height. As will be seen from the illustration, it bristled with points, for the Bohemian wizard argued rightly that five points would be more efficient than one, and 50 more efficacious than five. The weird-looking structure destined to ward off the lightning of heaven had no less than 325 well-distributed points. Lodge says in his *Lightning Conductors*: "Points to the sky are recognized as correct; only I wish to advocate more of them, any number of them, like barbed wire along ridges and eaves. If you want to neutralize a thunder-bolt, three points are not as effective as 3000." This was written in 1892; nearly 140 years before that date, we find a simple parish priest of an obscure village in Moravia using precisely such a multiple system of short, pointed conductors for the protection of life and property. This lightning conductor or *meteorological machine*, as Divisch called it, was erected by him at Prenditz on June 15th, 1754. On the top of the rod will be seen three light vanes, which were added in the interest of the feathered race in order to prevent incautious members from incurring the risk of electrocution by alighting on the apparatus during a storm. The wind whirled the vanes round like the cups of an anemometer, and thus kept the birds away from the zone of danger.

Several trials came to the electrical Pastor, and from quarters least expected. It happened in the second year after the erection of the apparatus that the summer was unusually dry, in consequence of which the crops failed almost completely. The farmers of the neighborhood were always suspicious of the strange-looking mast of Prenditz; and, be it said, that they were more than diffident about the propriety of interfering with the forces of nature even under the plea of protection, forgetting that they took great care to protect themselves against heat and cold, rain, snow and hail. The country ladies, no doubt, used parasols for one kind of protection; and the gentry, umbrellas for another. Anyhow, the people of Prenditz and the good folk around did not like the lofty mast, with its outstretched arms and bristling rows of suspicious-looking iron points connected to the ground by means of four long, heavy chains. For the nonce, they deemed their Pastor a queer fellow, who thought that he could avert the anger of heaven by the oddest kind of a machine which they ever laid their eyes on. It was argued in the councils of the hamlets that, whatever advantages Divisch claimed for his "machine," they were all of a negative character. It *prevented* the lightning stroke, he said; that might be, but they did not *see* the prevention. What they did see and keenly realize was the failure of their crops. That affected them very closely; and if, as they supposed, the apparatus of Prenditz had anything to do with it, the sooner they got rid of the machine the better. Divisch, it must be said, was liked by his people; but despite his popularity, the men of violence carried the day and the machine was doomed. Popular passion, excited by personal interest, got the better of the consideration due to the Pastor. On an appointed day, a band of bellicose farmers came down on the village and wrecked the apparatus which had cost the priest so much thought and manual labor and on which, knowingly and justly, he relied for the protection of the homesteads of his rustic flock.

This recalls a similar incident of mob violence which occurred at St. Omer in the north of France, where a manufacturer of that quaint old town, who had been in America and seen the usefulness of lightning conductors, proceeded to erect one over his own house. Hardly was it completed before the populace gathered together; and, when passion was sufficiently aroused by inflammatory remarks of the demagogues, the house was attacked and the conductor torn down. The manufacturer complained of the inaction of the "gardiens de la paix" and appealed to the courts to uphold his right to protect his home against lightning. He entrusted his case to a young, brilliant lawyer, as yet unknown to fame, but one destined to achieve unenviable notoriety during the revolutionary period. This, the first defender of the lightning-rod in a court of justice, was Robespierre.

The news of the untoward event soon reached the ears of the Premonstratensian's superiors at Kloster-Bruck; and, as they very wisely considered that the duty of a country priest is primarily to attend to the spiritual welfare of his people, rather than to invent machines for their protection against the bolts of heaven, they advised him to yield to the prejudice of his people and not reconstruct the objectionable apparatus.

Father Divisch accepted the friendly advice of his superiors and obeyed like a good Premonstratensian monk. The remains of the shattered "meteorological machine" were sent to the abbey at Bruck, where they could be seen for many years afterward. As a consequence of this act of vandalism, Divisch gave up experimenting with lightning-rods and with electricity itself. The villagers were satisfied, but the world at large lost the benefit that might accrue from the researches on atmospheric electricity which Divisch would have carried on during the remaining nineteen years of his life.

In giving up electricity, the disappointed priest turned his attention, first, to acoustics and then, practical man as he was, to the construction of musical instruments. It was not long before his genius brought out an orchestrion of wind and stringed instruments which was played like an organ with hands and feet, and which was capable of 130 different combinations. Prince Henry of Prussia offered a considerable sum of money for the invention, but Divisch died while the preliminaries of sale were arranging, and negotiations were broken off. The instrument remained for many years in the abbey at Bruck, where it was in daily use for the canonical office.

It is a curious coincidence that Franklin was also interested in musical instruments. He is credited with having devised an improved form of glass harmonica, one of which he presented to Queen Marie Antoinette.

Despite the bitter experience of Divisch, the introduction of lightning conductors into Italy was warmly advocated some years later by Padre Toaldo (1719-1797), an admirer and correspondent of Franklin. It was through his influence and personal activity that the magnificent thirteenth-century Cathedral of Siena was protected with lightning conductors after having been repeatedly struck during the centuries and seriously damaged. Toaldo published in 1774 his celebrated work on the protection of public edifices and private buildings against lightning; it contributed greatly to reassure public opinion on the value of "Franklinian rods," as the conductors were commonly called.

It is a matter of regret that Franklin used the words "the electric fluid is attracted by the points" in the passage quoted above, inasmuch as in the popular mind such "attraction" courts rather than averts danger. As already said, the rod no more "attracts" lightning than a rain-pipe attracts a downpour. Franklin knew very well the twofold function of his rods, the *preventive,* by which they tend to ward off the stroke by gradually and silently neutralizing the excessive energy of the cloud; and the other, the *preservative,* by which they convey the discharge safely to earth when struck. He even complains of people who concentrate their attention on the preventive function, forgetting the other entirely, adding that, "Wherever my opinion is examined in Europe, nothing is considered but the probability of these rods preventing a stroke, which is only a part of the use which I proposed for them; and the other part, their conducting a stroke which they may happen not to prevent, seems to be totally forgotten, though of equal importance and advantage." (1755.)

At a time, it was customary to make the rods rise to a considerable height above the building, in the belief that the diameter of the circle of protection was four times the height of the rod. Such a rule was an arbitrary one which facts soon showed to be unreliable and unsafe. It is now recognized that there is no such thing as a definite area of protection.

Were this a literary chapter, we would point out that either of the expressions "electric" storm or "lightning" storm is preferable to *thunder-storm,* because electricity or lightning is the active agent or principal feature of the impressive phenomenon. No one thinks of calling a hailstorm by the descriptive term of *patter-storm;* yet that would be just as logical and appropriate an appellative in one case as thunder-storm is in the other.

Thunder-tube is certainly a startling misnomer applied to the long, narrow, glazed tubes formed in siliceous materials by the fervid heat of the flash, but not in any way by the sound-waves produced by the crash. *Thunder-bolt* does not mean, despite the common opinion, a white-hot mass that accompanies the discharge; it is purely and simply the flash itself. A glowing mass that happens to come down in the track of the discharge is a *meteorite,* a body of cosmic not terrestrial origin, a visitor from space that chose the rarefied path of the flash for its descent to earth.

Again, there are no *thunder-clouds* in nature, only electric clouds or lightning clouds; nor is there ever *thunder in the air* save when the lightning breaks from cloud to cloud, or leaps from cloud to earth, or strikes from earth to cloud. But though thunder is only occasionally in the air, electricity always is. We have a normal electrical field in all seasons, times and places.

Though it is the lightning that kills and not the thunder, we would not, however, object to the following inscription which we found on a tombstone:

"Here lies (so and so), oh! what a wonder, She was killed outright by a peal of thunder,"

because the suddenness of the peal may have given the aged lady a shock from which her failing heart was unable to recover.

We are well aware that such criticism of technical terms in popular use will have no reform effect whatever; because as long as people will say "the sun rises" and "the stars set," they will continue to speak of thunder-clouds and thunder-storms, thunder-tubes and thunder-bolts. Though containing an element of error, these expressions have the sanction of the centuries; and so, they have come to stay.

Returning to Divisch, that worthy priest and pioneer electrician died at Prenditz in his sixty-ninth year, on Dec. 21st, 1765, and was buried in the little churchyard where he had blessed many a grave during the twenty-five years of his ministration. A simple inscription marks the place of his interment, but a monument will soon be erected to his memory which will tell the passerby where sleeps the Premonstratensian pioneer of the lightning-rod.

About three months before the erection of his rod, *i.e.*, in June, 1752, the idea occurred to Franklin that he could approach the region of clouds just as well by means of a common kite. Here are his words anent the novel and famous experiment with the "lightning kite":

"Make a small cross of two light strips of cedar, the arms so long as to reach to the four corners of a large thin silk handkerchief when extended; tie the corners of the handkerchief to the extremities of the cross, so you have the body of a kite, which, being properly accommodated with a tail, loop and string, will rise in the air, like those made of paper; but this, being of silk, is fitter to bear the wet and wind of a thunder-gust without tearing. To the top of the upright stick is to be fixed a very sharp-pointed wire, rising a foot or two above the wood. In the end of the twine, next the hand, is to be held a silk ribbon, and where the silk and cord join a key may be fastened. This kite is to be raised when a thunder-gust appears to be coming on, and the person who holds the string must stand within a door or window, or under some cover, so that the silk ribbon may not be wet; and care must be taken that the twine does not touch the frame of the door or window. As soon as any of the thunder-clouds come over the kite, the pointed wire will draw the electric fire from them, and the kite with all the twine will be electrified, and the loose filaments of the twine will stand out every way and be attracted by an approaching finger. And when the rain has wetted the kite, so that it can conduct the electric fire freely, you will find it stream out plentifully from the key on the approach of your knuckle. At this key the phial may be charged, and from electric fire thus obtained spirits may be kindled and all the other electric experiments be performed which are usually done by the help of a rubbed glass globe or tube, and thereby the sameness of the electric matter with that of lightning completely demonstrated."

Here we have the electric kite and manner of using it fully described without, however, any direct statement that the author himself actually experimented with it, although he does say that the experiment was successfully carried out. This is strictly true, but it may be safely contended that the precautions enumerated, the observation about the fibres of the cord, its improved conductivity when wetted by the rain and the like, all bespeak a knowledge of practical conditions that could be obtained only by way of experiment.

But if Franklin is not outspoken on the matter, some of his contemporaries are. Here is the kite incident as related in the *Continuation of the Life of Dr. Franklin*, by Dr. Stuber, a Philadelphian and intimate friend of the Franklins:

"While Franklin was waiting for the erection of a spire, it occurred to him that he might have more ready access to the region of clouds by means of a common kite. He prepared one by fastening two cross-sticks to a silk handkerchief, which would not suffer so much from the rain as paper. To the upright stick was affixed an iron point. The string was, as usual, of hemp, except the lower end, which was silk. Where the hempen string terminated, a key was fastened. With this apparatus, on the appearance of a thunder-gust approaching, he went out into the commons, accompanied by his son, to whom alone he communicated his intentions, well knowing the ridicule which, too generally for the interest of science, awaits unsuccessful experiments in philosophy. He placed himself under a shed to avoid the rain. His kite was raised. A thunder-cloud passed over it. No sign of electricity appeared. He almost despaired of success, when suddenly he observed the loose fibres of his string move toward an erect position. He now presented his knuckle to the key and received a strong spark. Repeated sparks were drawn from the key, the phial was charged, a shock given, and all the experiments made which are usually performed with electricity."

This testimony of a man who enjoyed the unlimited confidence of Franklin has a very matter-of-fact ring about it; there is not a note of uncertainty, not a word indicating doubt that his friend and neighbor went out to the fields accompanied by his robust son, carrying along with them a queer assortment of electrical impedimenta. This son, William by name, was twenty-two years of age at the time; and as he died in 1813, eleven years after the publication of Dr. Stuber's biographical sketch, he had ample time to contradict the kite story if instead of being a fact it were a mere romance. Nor is this all, for Dr. Stuber's narrative, given above, appears textually in the "Memoirs of the Life and Writings of Benjamin Franklin," edited by his grandson William Temple Franklin. The Doctor, be it remarked, was very fond of his grandson, whose "faithful service and filial attachment" he warmly commends in several of his letters, and whose regard for the memory of the statesman led him to undertake the task of preparing his works for publication. On page 211, Vol. I., he tells us that "As Dr. Franklin mentioned his electrical discoveries only in a very transient way, and as they are of a most important and interesting nature, it has been thought that a short disgression on the subject would be excusable and not void of entertainment. For this purpose the following account of the same, including the first experiment of the lightning kite, as given by Dr. Stuber, is here given."

In these concluding lines we have the testimony of Franklin's grandson to the authenticity of the "lightning kite" story. Moreover, the account as given by Stuber evidently meets with his cordial approval, since he transcribes it verbatim; and, as if to invest the quotations with unimpeachable authority, he tells us in the preface, p. viii., that "they deserve entire dependence because of the accuracy of the information imparted."

A word now from Priestley, also one of Franklin's intimate friends. In his *History of Electricity*, fourth edition, p. 171, he says that "Dr. Franklin, astonishing as it must have appeared, continued actually to bring lightning from the heavens by means of an electrical kite which he raised when a storm of thunder was perceived to be coming on." Then follows a description taken almost word for word from Dr. Stuber, whom he styles "the best authority on the subject."

If, perchance, the above testimony should not be deemed conclusive and final, all lingering doubt must be removed by Franklin's own words, for in his *Autobiography,* after briefly referring to the experiments made in France with pointed conductors, he adds: "I will not swell this narrative with an account of that capital experiment (the pointed conductor), nor of the infinite pleasure which I received on the success of a similar one I made soon after with a kite at Philadelphia, as both are to be found in histories of electricity."

Here, at last, we have Franklin's own word for it, that he made the kite experiment, and that he made it "soon after" the demonstration of his electrical discoveries which M. de Lor gave, by request, before Louis XV. and his court.

The "lightning kite" is, therefore, not a myth, as some have ventured to think, having been fully described by Franklin in his letter to Peter Collinson, dated October 19th, 1752, and having been made by him some time in June of the same year.

We have now to see whether Franklin was anticipated in the idea of the kite or in its use for electrical purposes. There are some who hold that he was anticipated by M. de Romas as to the idea, but not the actual experiment; while others credit the French magistrate with both. Let us examine the evidence which there is for these opinions.

M. de Romas lived in Nérac, a small town some seventy-five miles south of Bordeaux. He was a member of the bar; and at the time of the Franklinian furor in Europe was a judge of the district court. He took an interest in scientific matters quite unusual for men of his profession, proceeding, as soon as he had read of the efficiency of pointed conductors, to study their behavior for himself. His experiments met with surprising success, and were as much admired by the local savants as they were dreaded by the common folk. Letters containing his observations were regularly sent to the Academy of Bordeaux, where they were read with lively interest on account of their character and novelty. From the published *Actes* of that body we learn that the first kite used by de Romas was raised by him on May 14th, 1753. Disappointment, however, attended this attempt, no electrical manifestation being observed, although rain fell and wetted the hempen cord. The magistrate of Nérac attributed his failure to the resistance of the string; and, like a good electrician, surprisingly good for the time, determined to improve its conductivity by wrapping a fine copper wire round its entire length.

When this long and tedious operation was completed, he went out again to the fields on a stormy day, when, assisted by two of his friends, he raised the kite and soon got torrents of sparks from the wire-wound cord. This was on June 7th, 1753. The experiment was repeated from time to time, both for his own satisfaction and that of his assistants as well as for the entertainment of his ever-growing class of admiring spectators. Kites 7-1/2 ft. long and 3 ft. wide were raised 400 and even 550 ft. above ground when flashes nine feet long and an inch thick were drawn, so the account says, with the report of a pistol. The effect must have been truly spectacular. The kite was held by a silk ribbon fastened to the end of the hempen cord.

It is then a matter of history vouched for by the *Actes* of the Academy of Bordeaux that May 14th, 1753, is the day on which the first use of a kite for electrical purposes was made in France; on the other hand, it is to be remembered that Franklin flew his "lightning kite" in June, 1752, almost a year earlier. As far, then, as the *fact* is concerned, the Philadelphia philosopher was not anticipated by the Justice of Nérac.

From facts let us pass to writings. Franklin's letter to Collinson, in which he describes the electric kite, is dated October 19th, 1752, while that of M. de Romas, on which the claim for priority is founded, was addressed by him to the Academy of Bordeaux on July 12th, 1752, three months earlier. After a lengthy and interesting account of his experiments with pointed conductors, he concludes his communication as follows:

"C'est là, Monsieur, ce qu'il y a de plus important, car j'aurais bien d'autres particularités à vous communiquer; mais ma lettre, devenue d'une excessive longueur, m'avertit de finir. Je me réserve de mettre au jour la dernière (quoiquelle ne soit qu'un jeu d'enfant) lorsque je me serai assuré de la réussite par l'expérience que je me propose d'en faire et que je ne negligerai pas."

In English this would read: "Such, Sir, are the more important points which I have to communicate, and to which many others might be added, were it not for the excessive length of this letter, which warns me that it is time to bring it to a close. I will, however, give publicity to the last one of all (though it is only a child's plaything) as soon as I shall have assured myself of its success by an experiment which I have devised and which I shall not fail to make."

The words in brackets—"though it is only a child's plaything"—are all important, for it is on them and on them alone that the claim for priority has been put forth and maintained. It will be seen that the word kite (*cerf-volant*), does not occur in the letter, so that there can be no absolute certainty as to the nature of the *jeu d'enfant* which the author had in mind, though it is very likely that the kite was meant. In his *Mémoire sur les moyens de se garantir de la foudre dans les maisons*, he says, after describing some experiments that he had made with pointed rods: "Néanmoins toujours plein du désir d'augmenter le volume du feu électricque, il fallut chercher le moyen pour y parvenir. En conséquence, je me plongeai dans de nouvelles méditations. Enfin une demi-heure après, tout au plus, le cerf-volant des enfants se présenta tout à coup à mon esprit, et il me tardait de la mettre à l'épreuve. Par malheur, je n'en avais pas le temps." In English: "Being anxious to augment the quantity of electric fire, I began to think of some means to effect my purpose, and soon became quite absorbed with the subject. Not more than half an hour elapsed before the idea of the kite suddenly occurred to me, and I longed for an opportunity to try it; but unfortunately I had not sufficient leisure at the time." The work in which this passage occurs was published at Bordeaux in 1776, shortly after the death of the author. De Romas always maintained that he did not borrow the idea of the kite from any one, but that it occurred to him while pursuing his experiments with pointed conductors.

It must be admitted that de Romas could not have been acquainted with Franklin's performance of June, 1752, when he sent to the Bordeaux Academy his letter of July 12th, of the same year, for we cannot suppose that in an age of sailing vessels such news would cross the Atlantic and reach an obscure provincial town in the southwest of France in the space of a month. On the other hand, it is equally improbable that a vague allusion to the electrical use of a kite made at Nérac on July 12th, by a man entirely unknown to fame as was de Romas, should be talked of on the banks of the Schuylkill before October 19th, the date of Franklin's memorable letter to Collinson. Moreover, the "*jeu d'enfant*" allusion as well as the very use of the kite by de Romas failed so completely to attract the attention of scientific men of his own country that he frequently and bitterly complained down to the end of his life, in 1776, of their persistent neglect of his claims to recognition.

From all this, we conclude:

(*a*) That Franklin's "lightning kite" is not a myth, the experiment having been made by him in June, 1752, and fully described by him in a memorable letter written to Peter Collinson, of London, dated October 19th of the same year:

(*b*) That de Romas independently had the idea of using a kite for electrical purposes as early as July 12th, 1752; but that he did not carry out his idea until May 14th, 1753; and, furthermore, that he did not succeed in getting any electrical manifestations until June 7th, 1753, his success then being due, at least in part, to the clever idea which he had of entwining the cord with a fine copper wire. Therefore, *suum cuique*.

In conclusion, we would say that the cardinal and enduring achievements of Franklin are:

(1) His rejection of the two-fluid theory of electricity and substitution of the one-fluid theory; (2) his coinage of the appropriate terms *positive* and *negative*, to denote an excess or a deficit of the common electric fluid; (3) his explanation of the Leyden jar, and, notably, his recognition of the paramount role played by the glass or dielectric; (4) his experimental demonstration of the identity of lightning and electricity; and (5) his invention of the lightning conductor for the protection of life and property, together with his clear statement of its preventive and protective functions.

If Franklin was well acquainted with electrical phenomena, it is safe to say that his knowledge of human nature was wider and deeper still. This appears continually in his *Autobiography*, in his political writings, in business transactions and diplomatic relations.

On one occasion, while his re-election as clerk of the General Assembly was pending, a certain member made a long speech against him. Franklin listened with calm, dignified composure; and after his election, instead of resenting the opposition of the offending member, he determined that it would be better to disarm his antagonism and win his friendship. For this purpose he sent the assemblyman a courteously-worded request for the loan of a very scarce book which was in his library. The book was sent to Franklin, who returned it within a week with a note of thanks, which had the desired effect. Commenting on the event, our philosopher says that "it is more profitable to remove than to resent inimical proceedings."

Some of Franklin's views on general political economy are tersely set forth in the following passage: "There seem, in fine, to be but three ways for a nation to acquire wealth. The first is by *war*, as the Romans did in plundering their conquered neighbor; this is *robbery*. The second is by *commerce*, which is generally *cheating*. The third is by *agriculture*, the only *honest way* wherein man receives a real increase of the seed thrown into the ground, in a kind of continual miracle wrought by the hand of God in his favour, as a reward for his innocent life and virtuous industry."

Franklin asserts his religious convictions in many passages of his "Autobiography" as well as on many occasions of his public life. Shocked by "Tom" Paine's views of fundamental religious truths, he says: "I have read your manuscript with some attention. By the argument which it contains against a particular Providence, though you allow a general Providence, you strike at the foundation of all religion. For, without the belief of a Providence that takes cognizance of, guards and guides, and may favour particular persons, there is no motive to worship a Deity, to fear His displeasure, or to pray for His protection. I will not enter into any discussion of your principles, though you seem to desire it. At present, I shall only give you my opinion that, though your reasonings are very subtile and may prevail with some readers, you will not succeed so as to change the general sentiments of mankind on that subject; and the consequence of printing this piece will be a great deal of odium drawn upon yourself, mischief to you, and no benefit to others. He that spits against the wind, spits in his own face."

This aphorism recalls the ripe wisdom contained in many of the sayings of "Poor Richard," for Franklin was a deep thinker, shrewd observer and quaint expositor of his own philosophy. Continuing, he fleeces Paine in the following noble words: "But were you to succeed, do you imagine any good would be done by it? You yourself may find it easy to live a virtuous life without the assistance afforded by religion; you having a clear perception of the advantages of virtue and the disadvantages of vice, and possessing strength of resolution sufficient to enable you to resist common temptations. But think how great a portion of mankind consists of weak and ignorant men and women, and of inexperienced, inconsiderate youth of both sexes, who have need of the motives of religion to restrain them from vice, to support them to virtue, and retain them in the practice of it till it becomes *habitual*, which is the great point for its security. And perhaps you are indebted to her originally, that is, to your religious education for the habits of virtue upon which you now justly value yourself. You might easily display your excellent talents of reasoning upon a less hazardous subject, and thereby obtain a rank with our most distinguished authors. For among us, it is not necessary, as among the Hottentots, that a youth, to be raised into the company of men, should prove his manhood by beating his mother."

Franklin concludes this magnificent expression of his religious faith by the solemn warning: "I would advise you, therefore, not to attempt unchaining the tiger, but to burn this piece before it is seen by any other person; whereby you will save yourself a great deal of mortification by the enemies it may raise against you, and perhaps a good deal of regret and repentance. If men are so wicked *with* religion, what would they be *without* it?"

Franklin's belief in the cardinal doctrine of the resurrection of the body is well expressed in the epitaph which he wrote for himself in 1728, when in his twenty-second year. It reads

The Body
Of
Benjamin Franklin
Printer,
(Like the cover of an old book
Its contents torn out
And stript of its lettering and gilding)
Lies here, food for worms.
But the work shall not be lost
For it will (as he believed) appear once more
In a new and more elegant edition
Revised and corrected
By
The Author.

However, when the statesman and philosopher was laid at rest beside his wife in the Cemetery of Christ Church, Philadelphia, in 1790, the marble slab which marked the grave bore no other inscription than Franklin's name and the date of his death.

Appreciating the great loss which the country sustained by the death of Franklin, Congress ordered a general mourning for one month throughout the fourteen States of the Union; and the French National Assembly decreed three days of public mourning at the instance of Mirabeau, who said in his address that "The genius that gave freedom to America and scattered torrents of light upon Europe, has returned to the bosom of the Divinity. Antiquity would have erected altars to that mortal who for the advantage of the human race, embracing both heaven and earth in his vast mind, knew how to subdue both thunder and tyranny."

The fugitive apprentice boy of 1723 turned out to be one of the most esteemed and eminent Americans of his day. Of an even temper and well-balanced mind, he was plain in dress, simple in manner, easy of approach and friendly to all. The success which he achieved during his long career of eighty-five years, shows what may be done by seizing the opportunities which come to every one, by concentration of mind, application to duty and tenacity of purpose. He attained distinction in science, in letters, in diplomacy; he stood for good government and true liberty. His name is a household one in his own country, where monuments, institutions and cities will bear it down to posterity.

ADDENDA.

The Lightning Kite.

Fully described by Franklin in a letter to Peter Collinson, of London, dated October 19th, 1752.

Stuber in his "Continuation of the Life of Dr. Franklin," and Priestley in his "History of Electricity," affirm that Franklin made the experiment in June, 1752.

Franklin's son, William, never denied the story, although he figured in it as an active character.

William Temple Franklin, who prepared for publication his grandfather's works, gives the kite story almost verbatim from Stuber.

Finally, Franklin himself states that he made the experiment: Memoirs, Vol. I., p. 164.

Franklin and de Romas.

June, 1752: Franklin raises his kite in a field near Philadelphia.

July 12, 1752: Letter of de Romas to the Academy of Bordeaux, in which a probable reference is made to the kite as *un jeu d'enfant*.

October 19th, 1752: Franklin describes the "lightning kite" in a letter to Peter Collinson, of London.

May 14th, 1753: First use by de Romas of the electric kite in the fields around Nérac; no result.

June 7th, 1753: First success by de Romas with his electric kite.

Pointed Conductor.

Suggested by Franklin in letter to Peter Collinson, of London, dated July 29th, 1750.

D'Alibard, following Franklin's instructions, gets torrents of discharges from his iron rod 40 feet high at Marly, May 10th, 1752.

De Lor gets good results from his conductor 99 feet high, erected over his house in Paris, May 18th, 1752.

De Buffon succeeds with his rod on May 19th, 1752.

Franklin erected the first rod over his house in Philadelphia in September, 1752. It was made of iron with a sharp steel point rising seven or eight feet above the roof, the other end being sunk five feet in the ground. Franklin charged a Leyden Jar from his rod in April, 1753. Professor Richmann, of St. Petersburg, was killed by a flash from his apparatus on August 6th, 1753.

CHAPTER IV.GALVANI, DISCOVERER OF ANIMAL ELECTRICITY.

It is a well-known fact, often commented on in the history of medicine, that Harvey, the discoverer of the circulation of the blood, did not give the details of his discovery to the public for some twenty years after he had first reached it. The reason for his delay was twofold. With the characteristic patience of a real investigator in science, Harvey wanted to work out the details of his discovery for himself before giving it to the public, and wished to be sure of all he would have to say about it before committing it to print. He had not, as had indeed none of the really great discoverers in science, that intense desire for publicity which causes smaller men to rush into print with their embryonic discoveries, or oftener, their supposed discoveries, the moment they get their first distant glimpse of a new truth or see some mirage of a distant scientific principle, perhaps already well known, in their heated imaginations. Small men squabble about priority in small discoveries, and rush headlong into print, lest some one should anticipate their wonderful observation. The example of Harvey can scarcely be commended too highly, for if followed, it would save the world of science a lot of bother and obviate the necessity of taking back many things that have been proclaimed in the name of science. Fortunately, it has been the rule among genuine students of science, not because of any deliberate imitation of their great predecessors, but because of modest assurance of the worth of their work and honest desire to perfect it before giving it to the world.

Luigi, or, as he preferred to be known himself, Aloysio Galvani, for the young prince of the house of Gonzaga whose canonization made him St. Aloysius was his patron in baptism and a favorite in life, presents an interesting exemplification of this characteristic trait of the really great discoverer in science, to wait calmly and work faithfully for thorough confirmation of his views before publishing them. His admirable patience in reaching the real significance of his discovery before proclaiming the results of his investigations is only a typical illustration of the modest thorough scientist that he was. It used to be said that Galvani's discovery of the twitchings of the frog's legs, which led him to give himself to serious investigations into animal electricity, was made more or less by accident in 1786. His views on the subject of animal electricity were not formally published until the appearance of his treatise, De Viribus Electricitatis in Motu Musculari Commentarius, in the eighth volume of the Memoirs of the Institute of Science of Bologna, published in 1791. This would seem to indicate that only five years elapsed between his original observation and the publication of his views. Even this interval may seem long enough to our modern notions of at least supposed rapidity of scientific progress, but we know now, from documents in the possession of the Institute of Science at Bologna, that, twenty years previous to the publication of this commentary, Galvani was deeply interested in the action of electricity upon the muscles of frogs, and was diligently and fruitfully occupied during his spare time with investigations upon this subject.

When, in Makers of Modern Medicine, I called special attention to the fact that practically all of the greatest discoverers in medicine had made their cardinal discovery, or at least the far-reaching observation that opened up for them the special career in investigation that was to make them famous, before they were thirty-five, one of my critics doubted the assertion and suggested the case of Galvani as a distinct exception. Ordinarily, it is presumed that his discovery of the twitchings of frogs' legs under the influence of electricity was made in 1786, when he was in his forty-ninth year. As a matter of fact, however, his first observations were made and his attention attracted to the importance of the subject when he was scarcely more than thirty. His career is indeed a striking example of the earliness in life at which a great man's work is likely to come to him, and yet illustrates very aptly the patience with which he devotes himself to it, without seeking the idle reputation to be derived from immediate announcement, if he really has the true spirit of the scientific investigator.

Galvani began original work of a high order very early in his medical career. His graduation thesis on the human skeleton treated especially of the formation and development of bone, and attracted no little attention. It is noteworthy because of the breadth of view in it, for it touches on the various questions relative to osteology, from the standpoint of physics and chemistry, as well as medicine and surgery. It was sufficient to obtain for its author the place of lecturer in anatomy in the University of Bologna, besides the post of director of the teaching of anatomy in the Institute of Sciences, a subsidiary institution. Here, from the very beginning, Galvani's course was popular. He was not, as we note elsewhere, a fluent talker, but he was one of the first who introduced experimental demonstrations of his subject into his lectures, and this made his teaching very attractive and drew crowds to his university courses.

Galvani's work as an anatomist, however, was done much more in comparative anatomy than in the study of the human being. He selected birds for the special subject of his first investigations in the field, and his monograph on the kidneys of birds attracted widespread attention among the scientists of Europe. As the farthest removed from man of the beings that are warm-blooded, these creatures have always attracted particular attention, and, quite apart from any interest in evolution, were the subject of special investigation. Owing to the facility with which they can be studied in embryonic stages in the hatching egg, most of the peculiarities of their structure and development are very well known now. The kidneys of the bird are especially interesting, because they represent a different phase of development from that of human beings. Galvani had selected, then, one of the cardinal or turning-point subjects in comparative anatomy. As he pointed out very clearly, the kidneys of birds differ very much among themselves, and the intense muscular action of this creature makes a large amount of excretory material, that must be disposed of, and consequently demands much more active kidney function than occurs in most other classes of animals. Galvani studied every feature—the vessels, the nerves, the canals—and almost necessarily pointed out many new points or added hitherto unknown details.

He next devoted himself to the study of the ear of the bird. This might seem to be of little special interest, since hearing is not one of the most characteristic qualities of the winged species. It so happens, however, that the semi-circular canals which are closely connected with the auditory apparatus in all animals are extremely large in birds. As a consequence of this, the avian auditory structures assume an importance in comparative anatomy quite like that of the kidneys in the same species. After Galvani had completed his studies, he found that he had been anticipated by another great Italian anatomist of the time, Antonio Scarpa (of Scarpa's triangle in human anatomy), who afterwards became the Chief Surgeon to Napoleon. Galvani abandoned the idea of publishing his book then, but published a short article, in which he added much to Scarpa's details and conclusions. His additions were particularly with regard to the semi-circular canals, which are probably the organ of direction, the necessity for which, in this species, for the purpose of flying, is so easy to understand. He also described with great care the single ossicle or small bone, which replaces the chain of little bones that exist in mammal ears, and pointed out that the shape of this bone and its appendages enabled it to fulfil, though single, all the functions of the hammer, the anvil and the stirrup bones in human beings.

Galvani's careful study of the semi-circular canals of various species of birds can perhaps be better appreciated from the fact that he made it a point to measure their size exactly, as compared to the semi-circular canals of most other creatures. He found that the semi-circular canals of the hawk, for instance, were larger than the corresponding structures in man or even in the cow or the horse. As these latter animals are many hundred times larger than the largest birds, the special significance of the canals in birds becomes manifest. In certain of the birds, as he pointed out, these structures are not semi-circles, nor indeed of circular form at all, but take on much more the shape of an ellipse, and, indeed, sometimes the arc of curvature of the ellipse is quite acute. He seems to have had no hint, however, of the function that we have in modern times assigned to these structures, that of presiding over direction and equilibrium, and discusses in his rather vigorous Latin what the physiological significance of them may be as regards hearing. He thinks that they add something to the acuity of hearing, and would seem to imply that in birds flying rapidly through the air, there was the necessity for a more perfect hearing apparatus than among other creatures, and that this was the reason for the huge development of their semicircular canals.

At this time the science of comparative anatomy was just beginning to attract widespread attention. John Hunter, in London, was doing a great work in this line, which placed him in the front rank of contributors to biology and collectors of important facts in all the sciences allied to anatomy and physiology. Galvani's work on birds, then, made him a pioneer in the biological sciences that were to attract so much attention during the nineteenth century. His experimental work in comparative anatomy, strange as it might seem, and apparently not to be expected, led him into the domain of electricity, through the observation of certain phenomena of animal electricity and the effects of electrical currents on animals.

Like so many other great discoveries in science, Galvani's first attraction to his subject of animal electricity is often said to have been the result of a happy accident. Of course it is easy to talk of accidents in these cases. Archimedes and his bath; the fall of the apple for Newton; Laennec's observation of the boys tapping on a log in the courtyard of the Louvre and the ready conduction of sound, from which he got his idea for the invention of the stethoscope; Lord Kelvin's eye-glass falling and showing him how a weightless arm for his electrometer might be obtained in a beam of light,—may all be called happy accidents if you will. Without the inventive scientific genius ready to take advantage of them, however, these accidents would not have been raised to the higher plane of important incidents in the history of science. These phenomena had probably occurred under men's eyes hundreds of times before, but there was no great mind ready to receive the seeds of thought suggested, nor to follow out the conclusions so obviously indicated. Galvani's observation of the twitching of the muscles of the frog under the influence of electricity, may be called one of the happy accidents of scientific development, but it was Galvani's own genius that made the accident happy.

There are two stories told as to the method of the first observation in this matter. Both of them make his wife an important factor in the discovery. According to a popular but less authentic form of the history, Galvani was engaged in preparing some frogs' legs as a special dainty for his wife, who was ill and liked this delicacy very much. He thought so much of her that he was doing this himself, in the hope that she would be thus more readily tempted to eat them. While so engaged, he exposed the large nerve of the animal's hind legs, and at the same time split the skin covering the muscles. In doing this he touched the nerve muscle preparation, as this has come to be called, with the scalpel and the forceps simultaneously, with the result that twitchings occurred. While seeking the cause of these twitchings, the idea of animal electricity came to him.

The other form of the story is told a little later in Galvani's own words in the analysis of his monograph on animal electricity. He does not mention his wife in it, but there is a tradition that she was present in the laboratory when the phenomenon of the twitching of the frog's legs was first noticed, and indeed that it was she who called his attention to the curious occurrence.

She was a woman of well-developed intellect, and her association with her father and also with her husband made her well acquainted with the anatomy and physiology of the day. She realized that what had occurred was quite out of the ordinary. She is even said to have suggested their possible connection with the presence and action of the electric apparatus. Husband and wife, then, together, by means of a series of observations determined that, whenever the apparatus was not in use the phenomenon of the convulsive movements of the frog's legs did not take place, notwithstanding irritation by the scalpel. Whenever the electric apparatus was working, however, then the phenomenon in question always took place. According to either form of the story, if we accept the traditions in the matter, Madame Galvani had an important part in the discovery.

Galvani's most important contribution to science is undoubtedly his De Viribus Electricitatis in Motu Musculari Commentarius—Commentary on the Forces of Electricity in Their Relation to Muscular Motion. Like many another epoch-making contribution to science, it is not a large work, but in his collected works in the edition of 1841, occupies altogether sixty-four pages, of scarcely more than two hundred and fifty words to the page. There are probably not more than fifteen thousand words in it altogether. It was published originally in the eighth volume of the Memoirs of the Institute of Science at Bologna, in 1791, but a reprint of it, with some modifications, was issued at Modena in the following year. This Modenese edition, published by the Societa Typographica, was annotated by Professor Giovanni Aldini, who also wrote an accompanying dissertation, De Animalis Electricae Theoriae Ortu Atque Incrementis, On the Rise and Development of the Theory of Animal Electricity. In this volume was also published a letter from Galvani to Professor Carminati, in Italian, on the Seat of Animal Electricity. These two editions are the sources to which we must turn for whatever Galvani tried to make known with regard to animal electricity.

This little volume consists of four parts: the first of which is devoted to a consideration of the effects of artificial electricity on muscular motion; the second is on the effect of atmospheric electricity on muscular motion; the third is on the effect of animal electricity on muscular motion; and the fourth consists of a series of conjectures and some conclusions from his observations. The arrangement of the work, as can readily be understood from this, is thoroughly scientific. Galvani proceeds from what was best known and most evident to what he knew less about, trying to enlarge the bounds of knowledge and then suggesting the conclusions that might be drawn from his work and offering a number of hints as to the possible significance of many of the phenomena that might form suggestive material for further experimentation along this same line. In spite of the forbiddingness of the Latin to a modern scientist, as a rule, the little work is well worthy of study because of its eminently scientific method and the excellent evidence it affords of the way serious students of science approached a scientific thesis before the beginning of the nineteenth century.

The first paragraph of this dissertation is of such fundamental significance, because it represents the primal work done in animal electricity, that it has seemed to me worth while presenting entire. The original Latin from which the translation is made, and from which a good idea of Galvani's Latin style may be obtained, is given in a note.

"I had dissected a frog and had prepared it, as in Figure 2 of the fifth plate (in which is shown a nerve muscle preparation), and had placed it upon a table on which there was an electric machine, while I set about doing certain other things. The frog was entirely separated from the conductor of the machine, and indeed was at no small distance away from it. While one of those who were assisting me touched lightly and by chance the point of his scalpel to the internal crural nerves of the frog, suddenly all the muscles of its limbs were seen to be so contracted that they seemed to have fallen into tonic convulsions. Another of my assistants, who was making ready to take up certain experiments in electricity with me, seemed to notice that this happened only at the moment when a spark came from the conductor of the machine. He was struck with the novelty of the phenomenon, and immediately spoke to me about it, for I was at the moment occupied with other things and mentally preoccupied. I was at once tempted to repeat the experiment, so as to make clear whatever might be obscure in it. For this purpose I took up the scalpel and moved its point close to one or the other of the crural nerves of the frog, while at the same time one of my assistants elicited sparks from the electric machine. The phenomenon happened exactly as before. Strong contractions took place in every muscle of the limb, and at the very moment when the sparks appeared, the animal was seized as it were with tetanus."

Galvani then explains in detail how he made observations on control frogs at moments when there were no electric sparks, and decided that the contact with the scalpel was only effective in producing twitchings when there was a simultaneous electric spark. He noted, also, that occasionally the contractions did not occur, in spite of the fulfilment of the conditions mentioned. He traced this to fatigue. He then proceeded to vary the experiment in many ways, decreasing the size of the scalpel, increasing and decreasing the size of the electric machine and varying the method of preparation of the frog, so as to decide just what the significance of the phenomenon was. In a general way, it may be said that this study shows Galvani as one of the most careful of experimentalists, though he has often been declared to be a theorizer, rather than an observer.

A very interesting anticipation of Galvani's original experiment, made long before his time by a great naturalist, the story of which serves to show that discoveries made before their time, that is, before people are ready to follow them up, fail to attract attention, has been called to my attention by Brother Potamian. In the second volume of the Dutch Naturalist Swammerdam's Works, page 839, is to be found the following passage: "Another experiment that is at once very curious and suggestive can be made if one separates the largest of the muscles of the thigh of the frog and so prepares it with its adherent nerve as to leave it unhurt. If after this has been done you take the tendons of this muscle, one in each hand, and irritate the hanging nerve by a little forceps or other instrument, the muscle will recover the former motion which it had lost. You will see at once that it contracts and that there will be an effort as it were to bring together the two hands which hold its tendons. This I demonstrated, in the year 1658, to the illustrious Duke of Tuscany then reigning, when he was at the moment in a state of mind that prompted him not to favor me. This same experiment can be repeated with the same muscle as often and for as long a time as any portion of the nerve remains uninjured, so that we may, therefore, irritate the muscle to its former contraction as often as we wish."

As a foundation classic in electricity, Galvani's De Viribus Electricitatis deserves more detailed analysis. The first part of the monograph is taken up with experiments of many kinds, with what may be called artificial sources of electricity—the electric machine, the Leyden jar, and other modes of electrical development. The second part treats of the effects of atmospheric electricity upon muscular motion, by which expression Galvani means lightning, though he also observed various electrical manifestations in the muscles of his frogs when there was no actual lightning but only darkening of the heavens, without actual passage of the current flash from one cloud to another or from the clouds to the earth. In this matter, Galvani displayed quite as much courage as patient observation. He knew the fate of Richmann, the Russian scientist, who had been struck dead by a lightning-bolt while making experiments not very different, yet he dared to place a lightning conductor on the highest point of his house, and to this conductor he attached a wire, which ran down to his laboratory. During a storm, he suspended on this metallic circuit, by means of their sciatic nerves, frogs' legs and the legs of other animals prepared for the purpose. To the feet of the animals he attached another Wire sufficiently long to reach down to the bottom of a well, thus grounding the circuit.

Not satisfied with this study of the influence of lightning and large electrical disturbances in the air on the preparation of the frog as he had made it, Galvani set about discovering whether even the slight differences in electrical potential which occur during the day in atmospheric electricity might not give rise, even in fair weather, to certain contractions of the frog's muscles. He made his observations for many days at many different hours and under varying conditions of light and shade, of heat and cold, without finding anything. There were occasional contractions, but they bore no definite relation to variations in the atmosphere, or the electric state of the atmosphere. Galvani satisfied himself of this very thoroughly, and with a patience and diligence worthy of emulation by a Fellow at a modern university working on a foundation for the determination of a particular question.

The third part of the work is the most important as well as the longest, and contains the ideas which are original with Galvani, but which met most opposition in his time and have only been properly appreciated in recent years. Galvani came to the conclusion that there is such a thing as animal electricity. This led to a famous controversy with Volta, in which their contemporaries judged that Galvani had the worst of it; but, as so often happens, their successors a century later would judge that Galvani's views were more in accord with what we know at the present time. Criticism is always easier than scientific advance, and in a controversy it is usually the man who writes most forcibly, rather than the one who thinks most deeply, who secures the assent of readers. This makes controversy in matters of science always unfortunate, for it does much more to retard than to help scientific progress.

Galvani insists, at the end of this chapter on animal electricity, that what he writes is entirely the result of experiment, and that he has tried in every way to make his experiments from a thoroughly critical standpoint. Those who repeat his observations will find this to be true, though he confesses that there are times when conditions not well understood seem to hinder the results that he usually obtained.

The fourth part of his commentary is taken up with certain conjectures, as he calls them, and some conclusions from his work. In this he suggests the use of electricity for the cure of certain nervous diseases, and especially for the treatment of the various forms of paralysis. The use of electricity for these cases had been previously suggested, and Bertholinus had told the story of patients who were utterly unable to move and who had recovered after having been in the neighborhood where a lightning-bolt had struck. To the minds of physicians of that time, this must have seemed proof positive of the curative value of lightning, and, therefore, of electricity, for paralytic conditions. The remedy was heroic, if not indeed positively risky, but its good effect could not be doubted. Unfortunately, as is always true in medical matters, the real question at issue in these cases is not so much the value of the remedy as the propriety of the diagnosis. Paralysis, in the sense of inability to use one or more limbs, may be due to many causes. There are a number of forms of functional or hysterical palsy, that is, of incapacity to use certain groups of muscles not dependent on any organic lesion, but upon some curious state of the nervous system which may pass away entirely, and which, indeed, seem to be dependent on the patient's state of mind. A number of so-called paralytic patients were cured by the earthquake in San Francisco; some are made to do the apparently impossible every year; they get up and walk because of the shock due to a fire or burglars. We know now that the electrical status of the individual is very carefully protected from disturbance by external electrical forces. What Galvani began has borne fruit in diagnosis more than treatment, so that his prophecy has been amply fulfilled. "The application of this method may throw light on the subject and experience may help us to understand it."

Among his conclusions, Galvani hints that electricity may not only proceed from the clouds during electrical disturbances, but also may proceed from the earth itself, and that living beings may be affected by this. He suggests, therefore, that plants and animals may be influenced in their growth and in their health by such electrical changes. He adds the suggestion that there may be some intimate connection between electrical phenomena and earthquakes, and suggests that, in countries where earthquakes are frequent, observations should be made by means of frogs' limbs in order to see whether there may not be some definite change in the electrical conditions of the atmosphere before and during the earthquake. He seems to have had some idea that the curious feelings which at times come before an earthquake to human beings, though they seem even more noticeable in animals, may be due to this change in atmospheric electricity.

We are rather prone to think that news of scientific discoveries traveled slowly in Europe in the eighteenth century. There is abundant evidence of the contrary in these sketches of electricians, and Galvani's case is one of the most striking. How much attention Galvani's discovery attracted and how soon definite details of it spread to the other end of Europe may be judged from the fact that, in 1793, Mr. Richard Fowler published a small book at Edinburgh bearing the title, Experiments and Observations Relative to the Influence Lately Discovered by M. Galvani, and commonly called Animal Electricity. This little book, which may be seen at the

Surgeons General Library, Washington, and in the Library of the American Institute of Electrical Engineers, New York, details a large number of experiments that Fowler had made during the preceding year or more, so that Galvani's work must have reached him within a few months after its publication. Fowler mentions the fact that Galvani had been occupied many years before this in the study of electric fishes, especially the *torpedo*, the *gymnotus electricus* and *silurus electricus*. He also mentions a curious observation of Cotugno, who, a few years before, had received a shock from a mouse while dissecting the little animal, which makes it clear that imagination played a role in helping to the introduction of the newer ideas with regard to animal electricity.

But before his discovery was to attract so much attention, Galvani had to work it out, and this is the merit of the man.

It is almost needless to say, these experiments upon frogs were not accomplished in a few days or a few weeks. Galvani had his duties as Professor of Anatomy to attend to besides the obligations imposed upon him as a busy practitioner of medicine and surgery. At that time, it was not nearly so much the custom as it is at the present, to use frogs for experiments, with the idea that conclusions might be obtained of value for the biological sciences generally, and especially for medicine. There has always been such an undercurrent of feeling, that such experiments have been more or less a beating of the air. Galvani found this opposition not only to his views with regard to animal electricity as enunciated after experimental demonstration, but also met with no little ridicule because of the supposed waste of time at occupations that could not be expected to lead to any practical results. It was the custom of scientific men to laugh somewhat scornfully at his patient persistence in studying out every detail of electrical action on the frog, and one of the supposedly prominent scientists of the time even dubbed him "the frog dancing master." This did not, however, deter Galvani from his work, though some of the bitter things must have proved cutting enough, and might have discouraged a smaller man, less confident of the scientific value of the work that he was doing.

His relations with his patients—for during all of his career he continued to practice, especially surgery and obstetrics—were of the friendliest character. While his distinction as a professor at the University gave him many opportunities for practice among the rich, he was always ready and willing to help the poor, and, indeed, seemed to feel more at home among poor patients than in the society of the wealthy and the noble. Even toward the end of his life, when the loss of many friends, and especially his wife, made him retire within himself much more than before, he continued to exercise his professional skill for the benefit of the poor, though he often refused to take cases that might have proved sources of considerable gain to him. Early in life, when he was very busy between his professorial work and his practice, he remarked more than once, on refusing to take the cases of wealthy patients, that they had the money with which to obtain other physicians, while the poor did not, and he would prefer to keep some time for his services to them. When ailing and miserable toward the end of his life, he still continued his practice, and was especially ready to spend his time with the poor. He was dying himself, as one of his biographers says, when he got up from a sick bed to see a dying woman who sent for him.

He was one of the most popular professors that the University of Bologna ever had. He was not, in the ordinary sense of the word, an orator, but he was a born teacher. The source of the enthusiasm which he aroused in his hearers was undoubtedly his own love for teaching and the power it gave him to express even intricate problems in simple, straightforward language. More than any of his colleagues, he understood that experiments and demonstrations must be the real groundwork of the teaching of science. Accordingly, very few of his lectures were given without the aid of these material helps to attract attention. Besides, he was known to be one who delighted to answer questions, and was perfectly frank about the limitations of his knowledge whenever there was no real answer to be given to a question that had been proposed. Though an original discoverer of the first rank, he was extremely modest, particularly when talking about the details of his discoveries or subjects relating to them.

Galvani was not a good talker, though he seems to have been a good teacher. He had little of that facility which wins friends easily and enables a man to shine with a borrowed lustre of knowledge, often enough quite superficial. What he said was almost sure to have a very serious meaning. While there is no doubt that Galvani was a genius, in the sense that he was one of the precious few who take the step across the boundary of the unknown and make a path along which it is easy for others to follow in reaching hitherto trackless regions in human speculation, he also had what is undoubtedly the main element in talent, for he was possessed to a high degree of the faculty for hard work. For this he regulated the hours of his labor very carefully. Only thus could he have accomplished what he did. It must not be forgotten that he was teaching anatomy and obstetrics at the University of Bologna, and, surprising as it may seem, doing both these tasks well. He was besides accomplishing good work in comparative anatomy and physiology by original investigations of a high order. In spite of all this, which would seem occupation enough and more for any one man, he was able to keep up a rather demanding practice.

He did not have many friends, but those whom he admitted to his intimacy were bound to him with the proverbial hoops of steel. With two men in Bologna he spent most of his leisure. They were Dr. Julio Cæsare Cingari, a distinguished physician of the city, and the well-known astronomer who held the chair of astronomy at the University, Francisco Sacchetti. With these he passed many a pleasant hour, and week after week they met at one another's houses to discuss scientific questions and the lighter topics of the day. Galvani was thoroughly respected by all the members of the Faculty at Bologna, though he did not seek many friendships, and indeed probably would have more or less resented the intrusions of acquaintances, because of the time that it would take from him. He was a very retiring man, caring not at all for social things, and least of all for that personal fame which has been so well defined as the being known by those whom one does not know. His happiness in life came to him from his work and from his domestic relations. His wife was one of those marvelous women, rarer than they should be, one is tempted to say, who are enough interested in their husband's intellectual work to add to the zest of discovery in the discussion of it with them, and who yet realize that it is by minimizing the little worries of life that they can best help their husbands.

A very interesting phase of the Italian University life of that time is revealed in two important incidents of Galvani's university career. One of his professors—one, by the way, for whom he seems to have had a great deal of respect, and to whose lectures he devoted much attention, was Laura Caterina Maria Bassi, the distinguished woman Professor of Philosophy at the University of Bologna, about the middle of the eighteenth century. It is doubtless to her teaching that Galvani owes some of his thorough-going conservatism in philosophic speculation, a conservatism that was of great service to him later on in life, in the midst of the ultra-radical principles which became fashionable just before and during the French Revolution. Madame Bassi seems to have had her influence on him for good not only during his student career, but also later in life, for she was the wife of a prominent physician in Bologna, and Galvani was often in social contact with her during her years of connection with the University.

As might, perhaps, be expected, seeing that his own happy domestic life showed him that an educated woman might be the center of intellectual influence, Galvani seems to have had no spirit of opposition to even the highest education for women. This is very well illustrated by the first formal lecture in his course on anatomy at the University, which had for its subject the models for the teaching of anatomy that had been made by Madame Manzolini. In the early part of the eighteenth century, Madame Manzolini had been the Professor of Anatomy at the University of Bologna, and in order to make the teaching of this difficult subject easier and more definite, she modeled with great care and delicate attention to every detail, so that they imitated actual dissections of the human body very closely, a set of wax figures, which replaced the human body for demonstration purposes, at least at the beginning of the anatomical course.

Galvani, in taking up the work of lecturer in anatomy, appreciated how much such a set of models would serve to make the introduction to anatomical study easy, yet at the same time without diminishing its exactness, and accordingly introduced his students to Madame Manzolini's set of models in his very first lecture. At the time, not a few of the teachers of anatomy at the Italian universities were inclined to consider the use of these models as rather an effeminate proceeding. Galvani's lack of prejudice in the matter shows the readiness of the man to accept the best, wherever he found it, without regard to persons or feelings.

Galvani's personal character was very pleasant, yet rather grave and serious. His panegyrist, Professor Giuseppe Venturoli, in the eulogium of Galvani, delivered in the Public Academy of the Institute of Bologna (1802) within five years after Galvani's death, says that Galvani was far from that coldness or lack of interest which sometimes characterizes scientists in their social relations, and which, as he naïvely says, is sometimes praised and sometimes blamed by those who write about them. Another side of Galvani's character is more interesting. He was ready to do all in his power for the poor. He conducted his obstetrical clinic particularly with a liberal benevolence and charity that deserve to be mentioned. When it is considered how much time his teaching and his charity took from him, it is rather surprising to find that he had enough left to enable him to devote himself with so much success to the difficult tasks he set himself in research and to the time-taking labors of controversy, which occupied many years after the announcement of his discoveries.

The most striking proof of the thorough conscientiousness with which he faced the duties of life is to be found in his conduct after the establishment of the so-called Cis-Alpine Republic in Italy. This was a government established merely by force of arms, maintained through French influence, without the consent of the people, and a plain usurpation of the rights of the previous government. Galvani considered himself bound in duty to the authority under which he had lived all his previous life and to which he had sworn fealty.

When the University of Bologna was reorganized under the new government, the first requirement of all those who were made professors was that they should take the oath of allegiance to the new government. This he refused to do. His motives can be readily understood, and though practically all the other professors of the University had taken the oath, he did not consider that this freed him from his conscientious obligations in the matter.

Accordingly he was dropped from the roll of professors and deprived of the never very large salary which he had obtained from this chair. On this sum he had practically depended for his existence, and he began to suffer from want. While he had been a successful practitioner of medicine, especially of surgery, he had always been very liberal, and had spent large sums of money in demonstrations for his lectures and personal experimentation and in materials for the museums of the University. He began to suffer from actual want, and friends had to come to his assistance. He refused, however, to give up his scruples in the matter and accept the professorship which was still open to him. Finally, at the end of two years, influence was brought to bear on the new government, and Galvani was allowed to accept his chair in the University without taking the oath of allegiance. This tribute came too late, however, and within a short time after his restoration to his professorship he died.

Galvani's conduct in this affair is the key-note to his character and conduct through life. For him duty was the paramount word, and success meant the accomplishment of duty. For getting on in the world and material rewards he had no use unless they came as the consequence of duty fulfilled. His action in the matter of the University professorship has of course been much discussed by his biographers.

His eulogist, Professor Venturoli, whom we have already quoted, and whose eulogium is to be found in the complete edition of Galvani's works issued at Bologna in 1841, has much to say with regard to Galvani's religious sentiments.

He says: "The great founder in electricity was deeply religious, and his piety clothed a heart that was not less affectionate and sensitive to affection than it was intrepid and courageous. When called upon to take the civic oath in a formula involved in ambiguous words, he did not believe that he ought, on so serious an occasion, to permit himself anything but the clear and precise expression of his sentiments, full as they were of honesty and rectitude. Refusing to take advantage of the suggestion that he should modify the oath by some declaration apart from the prescribed formula, though it might still be generally understood that he had taken the oath, he refused constantly to commit himself to any such subterfuge. It is not our duty here to ask whether his conclusion was correct or not. He followed the voice of his conscience, which ever must be the standard of duty, and it certainly would have been a fault to have deviated from it. It is sad to think that this great man, deprived of his position, saw himself, for an instant at least, exposed to the danger of ending his career, deprived of the recompense which he so richly deserved and to which his past services to the State and the University had given him so just a title. This is all the more sad when we realize that the vicissitudes of his delicate health, much more than his age, now rendered such recompense doubly necessary. It is a gracious thing to recall, however, the noble firmness with which he maintained himself against so serious a blow. His courage is all the more admirable as one can see how absolutely without affectation it is. He was not ostentatious in his goodness, and did not permit himself to be cast down by the unfortunate conditions, but constantly preserved in the midst of adverse fortune that modest, imperturbable and dignified conduct which had always characterized him in the midst of his prosperity and his glory."

That his action in this matter was very properly appreciated by his contemporaries, and that the moral influence of his example was not lost, can be realized from the expressions used by Alibert, the Secretary-General of the Medical Society of Emulation, in the historical address on Galvani which he delivered before that society in Paris in 1801:

"Galvani constantly refused to take the civil oath demanded by the decrees of the Cis-Alpine Republic. Who can blame him for having followed the voice of his conscience—that sacred, interior voice which alone prescribes the duties of man and which has preceded all human laws? Who could not praise him for having sacrificed all such exemplary resignation, all the emoluments of his professorship, rather than violate the solemn engagements made under religious sanction?"

In the same panegyric there is a very curiously interesting passage with regard to Galvani's habit of frequently closing his lectures by calling attention to the complexity yet the purposefulness of natural things, and the inevitable conclusion that they must have been created with a definite purpose by a Supreme Being possessed of intelligence. At the time that Alibert wrote his memoir, it was the fashion to consider, at least in France, that Christianity was a thing of the past, and that while theism might remain, that would be all that could be expected to survive the crumbling effect of the emancipation of man.

He says: "We have seen already what was Galvani's zeal and his love for the religion which he professed. We may add that, in his public demonstrations, he never finished his lectures without exhorting his pupils to a renewal of their faith, by leading them always back to the idea of the eternal Providence which develops, preserves and causes life to flow among so many different kinds of things. I write now," he continues, "in the age of reason, of tolerance and of light. Must I then defend Galvani in the eyes of posterity for one of the most beautiful sentiments that can spring from the nature of man? No; and they are but little initiated in the saner mechanism of philosophy who refuse to recognize the truths established on evidence so strong and so authentic. *Breves haustus in philosophia ad atheismum ducunt, longiores autem reducunt ad Deum*—Small draughts of philosophy lead to atheism, but longer draughts bring one back to God"—(which may be better translated, perhaps, for English readers by Pope's well known lines, "A little learning [in philosophy] is a dangerous thing; drink deep or touch not the Pierian spring").

Galvani has been honored by his fellow-citizens of Bologna as one of their greatest townsmen, and by the University as one of her worthiest sons. In 1804, a medal was struck in his honor, on the reverse of which, surrounding a figure of the genius of science, were the two legends: "Mors mihi vita," "Death is life for me," and "Spiritus intus alit," "The spirit works within," which were favorite expressions of the great scientist while living, and are lively symbols of the spirit which animated him. In 1814, a monument was erected to him in the courtyard of the University of Bologna. It is surmounted by his bust, made by the most distinguished Bolognian sculptor of the time, De Maria. On the pedestal there are two figures in bas-relief, executed by the same sculptor, which represent religion and philosophy, the inspiring genius of Galvani's life.

Before he died, he asked, as had his favorite poet Dante, whose Divina Commedia had been one of the pleasures of life and above all one of the consolations of his times of adversity, to be buried in the humble habit of a member of the Third Order of St. Francis. He is said to have valued his fellowship with the sons of the "poor little man of Assisi" more than the many honorary fellowships of various kinds which had been conferred upon him by scientific societies all over Europe. With him passed away one of the great pioneers of modern science and one of the most lovable men in all the history of science. His death took place just before the close of the eighteen century, Dec. 4, 1798, but his work was destined to be one of the harbingers of a great period of electrical development.

CHAPTER V. VOLTA THE FOUNDER OF ELECTRICAL SCIENCE.

Up to the end of the eighteenth century, discoverers in electrical science had usually been students of science in other departments, whose attention to electricity had been attracted in passing as it were. Occasionally, indeed, they had been only interested amateurs, inquisitive as to the curious phenomena of magnetism. It is surprising how many of these pioneers in electricity were clergymen, though that fact is seldom realized. It can be seen very readily in my chapter on Clergymen Pioneers in Electricity, in Catholic Churchmen in Science (Second Series, Dolphin Press, Phila., 1909). With Volta's career, however, was initiated the story of the electrical scientists who devoted themselves almost exclusively to this department of physics, though more or less necessarily paying some attention to related subjects. Volta's discovery of a practical instrument for measuring electricity, as well as of comparatively simple apparatus producing a continuous current, changed the whole face of the science of electricity. After these inventions, regular work could be readily done in the investigation of problems in the science of electricity without discouragement or inadequate instruments, discontinuous electrical phenomena, disturbances of experiments by the weather, and other conditions which had been hitherto so unfavorable to electrical experimentation. Volta's invention of the pile, or battery, so deservedly called after him, caused electrical science to take on an entirely new aspect, and the modern development of electricity was assured. It has been well said that no other invention, not even the steam-engine, meant so much for the transformation of modern life as this new apparatus for the production of a continuous electric current.

Like most of the distinguished scientific discoverers of the last two centuries, Alessandro Volta was born in very humble circumstances. His father was a member of the Italian nobility, but had wasted his patrimony so completely that the family was in extreme poverty when the distinguished son was born, on the eighteenth of February, 1745. This poverty was so complete that Volta said of it, later in life:

"My father owned nothing except a small dwelling worth about fourteen thousand lire; and as he left behind him seventeen thousand lire of debt, I was actually poorer than poor." A good idea of the circumstances in which Volta's childhood was passed may be gathered from the fact that he could not even secure copy-books for his first school exercises except through the kindness of friends.

Volta had shown signs of genius from early boyhood, and yet had been discouragingly slow in his intellectual development as a child. In fact, it was feared that he was congenitally lacking in intelligence to a great degree. It is said that he was more than four years old before he ever uttered a word. This does not mean before he learned to talk connectedly, but before he could utter even such familiar expressions as "father," "mother," and the like. He was considered to be dumb; and, as is not infrequently the mistaken notion with regard to children dumb for any reason, he was thought to be almost an idiot. The first word he ever uttered is said to have been a vigorous "No!" which was heard when one of his relatives insisted on his doing something that he did not wish to do. At the age of seven, however, he had so far overcome all difficulties of speech as to be looked upon as a very bright child. Owing to this late, unexpected development, his parents seem to have regarded him as a sort of living miracle, and felt certain that he was destined to accomplish great things. His father said of him later, "We had a jewel in the house and did not know it."

Fortunately for Volta, one of his uncles was archdeacon of the Cathedral, and another was one of the canons. These relatives helped him to obtain an education, the way being made especially easy by the fact that at this time all the Jesuit Colleges subsisted on foundations and collected no fees from any of their students; so that all that was necessary for his uncles to do for him was to contribute to his expenses outside of college. According to tradition, the Jesuits not only helped Volta in his education, but assisted him in obtaining his books and even in his living expenses while at their college. At the age of about sixteen, his education was complete, even including a year of philosophy. This is probably an indication of his talent as a student; though it was not an unusual thing in the southern countries for students to graduate at sixteen, or even younger, after a course equivalent to that now required for the bachelor's degree in arts.

We have gotten far away from this early graduation, although it is still sometimes possible in Italian universities; and one of the brightest men I ever knew was an Italian who had graduated with a degree equivalent to our A. B. before he was sixteen. When Volta graduated, however, such early completion of the undergraduate course was not at all unusual in Italy, and boys of thirteen and fourteen, almost as a rule, entered the undergraduate department to complete their course for a degree at seventeen or eighteen. One of our greatest physicians in this country, Benjamin Rush, was only seventeen when he completed his college course, and such examples were not at all rare. Indeed, the possibility for these men to devote themselves much earlier than is possible now to their serious life-work, yet with the development of mind which comes from a University course in the arts, was probably a distinct help to the success of their scientific careers. One is tempted to think that possibly such justification of earlier graduation, as we find among the distinguished scientists of a century ago, might make us reflect deeply before lending ourselves to what Herbert Spencer thought a phase of evolution, the lengthening of childhood, for it is just possible that the earlier recognition of manhood may mean more for individual development. Of course, geniuses are exceptions to rule, and an argument founded on their careers may mean very little for the generality of students.

Like many another of the great scientists, Volta was not that constant source of satisfaction to his teachers while at school that might possibly be expected. He had little interest in the conventional elementary education of the time, he was frequently distracted during school hours, and even as a mere boy often asked questions with regard to natural phenomena that were puzzlers to his masters, and sometimes complained of their lack of knowledge. He fortunately outgrew this priggishness, for in later childhood he seems to have been one of those talented children who learn rapidly and who are impatient at being kept back while their slower fellow-pupils are having drilled into them what came so easy to readier talents.

In his classical studies, however, Volta was deeply interested. He was especially enthusiastic over poetry, and at school devoted the spare time that his readiness of acquisition left him to the reading of Virgil and Tasso. These favorite authors became so familiar to him that he could repeat much of them by heart, and even in old age could cap verses from them better than any of his friends, even those all of whose lives had been devoted exclusively to literary occupations. During his walks, when an old man, he often entertained himself by repeating long passages from the classic Latin and Italian poets.

Even at this time, Volta's interest in the physical sciences was very marked. There is still extant a Latin poem of about five hundred verses, in which he sets forth the observations of Priestley, the discoverer of oxygen, whom it used to be the custom to call the Father of Modern Chemistry. This poem shows his thorough familiarity with the work of the great English investigator. Volta's model was Lucretius. Lest it should be a source of surprise that an Italian scientist had recourse to Latin for even a poetic account of scientific discoveries, it may be well to recall that Latin was still the universal language of science at that time, and Volta's great contemporary in electricity, Galvani, wrote his original monograph on animal electricity in that language, and even the Father of Pathology wrote his first great treatise, De Causis et Sedibus Morborum, in that tongue. As to his adoption of verse as a vehicle for scientific writing, it must not be forgotten that, at the time when Volta was writing his poem, another distinguished writer on scientific subjects, Erasmus Darwin, the grandfather of Charles Darwin of the last generation, was composing his "Zoonomia; or, Animal Biography," in English verse. Didactic verse was quite the fashion of the time, and some of it, even when it came from acknowledged poets, had not more poetry than Volta's effusion.

As if to make up for his lack of linguistic faculty when young, Volta seems to have had a special gift for languages when he grew older.

Before the age of twenty, he knew French as well as his mother tongue, read German and English fluently, and Low Dutch and Spanish were not beyond his comprehension. Besides his verses in Latin he wrote poetry also in French and Italian, always with cleverness at least, and at times with true poetic feeling.

While attending the Jesuit school, he expressed, it is said, a desire to enter the Order. As his father, however, had been with the Jesuits for eleven years and had then given up his studies, his family feared a repetition of such an experience; and so his clergymen uncles took him away from the school and sent him for a while to the Seminary at Benzi. After a time Volta abandoned the idea of becoming a priest, but would not consent to follow the wishes of the family council further, at least not to the extent of becoming a lawyer. Though he studied law for a time, he constantly wandered away to the reading of books on the natural sciences and to the study of natural objects. Finally he was allowed to give up law to devote himself exclusively to science.

Fortunately, one of the canons of the Cathedral of Como, a former fellow-student of his and a man of considerable means, was also interested in the natural sciences, and obtained the books and instruments necessary to enable Volta and himself to continue their studies. Father Gattoni seems to have realized at once the possibilities for great advances in science that lay in Volta's wonderful powers of observation, and encouraged him in every way. As a consequence, some of the important experiments that laid the foundation of the modern science of electricity and proved the beginning of Volta's world-wide reputation were carried on in Gattoni's rooms.

As a young man, Volta was so completely devoted to scientific investigations that there could be no doubt of the bent of his genius for original work of a high order. His power of concentration of attention on a subject was supreme. Biographers emphasize that there was no time, much less inclination, for the levities that so often appeal to the growing youth. He was almost too staid and preoccupied with his work for his own health and the comfort of his friends. When he became interested in a series of experiments, he often forgot the flight of time, and was known to miss meals, and inadvertently to put off going to bed—apparently quite unconscious of his physical necessities. This intense concentration of mind had its disadvantages. One of his friends complained playfully that he made a rather disagreeable traveling companion on account of his tendency to become abstracted; and on occasions this friend was deeply mortified to see Volta, when in company, take out a pocket-handkerchief that had been used for some purpose in the laboratory—which showed unmistakable signs of its previous employment as a cleansing agent for dirty instruments or hands, though its possessor was evidently unconscious of its appearance. More than once, too, his handkerchief proved, when taken out for its natural uses, to be as preoccupied as its owner: specimens of rocks or natural curiosities that he had gathered and inadvertently allowed to remain in his pocket came with it.

All during his life he retained an unusual faculty for concentrating his attention, which at times amounted to complete abstraction from his surroundings. It is related that, one cold morning his students at the University of Pavia found him in his shirt sleeves, so intent on arranging the experiments that were to illustrate his morning lecture that he was unconscious of the time, and even did not notice their coming into the room until they had been for some time in their seats and he had finally completed the arrangement for the demonstrations. He was constantly occupied with problems in natural science, looking for the explanation of phenomena that he did not understand as well as gathering new data by observation and experiment. He was gifted with the supremely inquisitive spirit, in the scientific sense of the epithet, and could not be satisfied with accepting things as he found them without knowing the reasons for them.

Volta furnishes another excellent illustration of how soon genius gets at its life-work. We have his own authority for the fact that he had come to certain conclusions with regard to the explanation of electrical phenomena, which, when he was only nineteen years of age, he set forth in a letter to the Abbé Nollet, who was then one of the best known experimenters and writers on electrical phenomena in Europe. Though so young, Volta had tried to simplify Franklin's theory of electricity by assuming that there was an action only between a (supposed) electrical substance and matter. It is curious to see how much he anticipated what was to be the thinking for more than a century after his time and practically down to the present day. He considers that all bodies, in their normal state, contain electricity in such proportion that electrical equilibrium is established within them. Electrical phenomena, then, are due to disturbances of this equilibrium. Such disturbances may be produced by physical means, as by friction or by chemical means, and even atmospheric electricity may be explained in the former way.

Volta's first formal paper on electricity, bearing the title *De Vi Attractiva Ignis Electrici*, was published in 1769, when he was twenty-four years of age. His second paper, *Novus Ac Simplicissimus, Electricorum Tentaminum Apparatus*—New and Very Simple. Apparatus for Electrical Tests, shows that Volta was getting beyond the stage of theorizing about electricity into the experimental work, which was to form the foundation of his contributions to electrical science. It is not surprising, then, that when he was just past thirty, in 1775, he was able to announce to Priestley his invention of the electrophorus. Priestley is usually thought of as one of the founders of modern chemistry, but he was known to his own generation, especially at this time, as the writer of a very interesting and complete history of electricity. It is characteristic of Volta's careful ways, that the reason for his letter to Priestley was in order to obtain information from him as to what extent this invention, which Volta knew, as far as he was concerned, to be original with himself, was novel in the domain of electrical advance.

With the intense interest in his work that we have noted, it is not surprising to find Volta's investigations proving fruitful. His active inventive genius stood him in good stead in enabling him to demonstrate principles by working instruments. The electrophorus is but one of the instruments that show the very practical character of the man. He was especially taken with the idea of securing some method of measuring electricity. Among other things, he invented the condensing electroscope, in which, instead of the ribbons of gold leaf now employed, he used straws. With this instrument he was able to demonstrate the presence of minute quantities of electricity developed under circumstances in which ordinarily the occurrence of any such phenomena would be unsuspected. These two instruments, the electroscope and the electrophorus, lifted the department of electricity out of the realm of theory into that of accurate scientific demonstration, and made the electrical departments of the physical laboratories of the time much more interesting and important than they had been before.

Though so early occupied with electricity, Volta did not confine himself to this subject, nor even to the wider field of physics, and that he did not hesitate, in his scientific inquisitiveness, to follow clues even in chemistry, is well illustrated by his first step in the investigation of gases. His attention being called to bubbles breaking on the surface of Lake Maggiore while on a fishing excursion, he set about finding their source, and noted that whenever the bottom of the lake near the shore was stirred somewhat a number of bubbles arose, and that the gas thus set free was inflammable. He constructed an electrical pistol in which gases thus set free were exploded by a spark from the electrophorus. About the same time, on the principle of the electrical pistol, he invented the eudiometer, an apparatus by means of which the oxygen content of air could be determined.

With regard to these inventions, Arago calls attention to a special quality that is peculiar to all of Volta's work. "There is not a single one of the discoveries of Professor Volta," says the distinguished French scientist, "which can be said to be the result of chance. Every instrument with which he has enriched science existed in principle in his imagination before an artisan began to put it into a material shape."

After these inventions and his previous work, it is not surprising that in 1774 Volta was offered the professorship of experimental physics in the College of Como. Here he labored for five years, until he received a call, in 1779, to the professorship of physics at the University of Pavia, where he was destined to remain in an active teaching capacity for a period of forty years.

Volta began his life-work as professor of physics at Pavia by extending his observations on gases. He was the first to demonstrate the expansion of gases under heat, especially as regards their increased expansibility at higher temperatures. Many observers had been at work on this problem before his time, but there were serious discrepancies in the results reported. Volta was the first to point out the reasons for the apparent inconsistencies of previous investigators' findings; and from his observations alone some valuable data might have been obtained for the establishment of what has since become known as the "law of Charles."

At this time, his knowledge of English enabled him to follow English discoveries closely, and he seems to have paid particular attention to the work of Cavendish and Priestley. Not long after Cavendish's description of the method of obtaining pure hydrogen, Volta made a series of observations on the relations of spongy platinum to this gas, and pointed out the spontaneous ignition that takes place when the two substances are brought together. This experiment is the basis of what has since been known as the hydrogen lamp, called, from the German observer who first made it a practical instrument, Dobereiner's lamp.

After seven years of teaching, Volta was given the opportunity to visit various parts of Europe, and took advantage of the occasion to meet most of the celebrated men of science. His linguistic faculty stood him in good stead during this sabbatical year, and his travel aided him in completing a thorough acquaintanceship with European languages as well as with scientists. His practical character led him, during his trip, to note the growth of the potato and its uses in various European countries, and he brought the plant home with him to Italy in order to introduce it among the farmers. He succeeded in making his countrymen realize its value, and the introduction of the potato is one of the reasons for which Italians have always looked up to him as a benefactor of his native land. How modern this makes a vegetable we are inclined to think of as having been always an important food resource of the race!

About the middle of the third quarter of the eighteenth century, by one of the fortunate accidents that happen, however, only to genius, Galvani, at the time Professor of anatomy in Bologna, had been led to make the observation that if a frog, so prepared that its hind leg is attached to the trunk only by means of the sciatic nerve, happens to be touched by a metal instrument in such a way as to put nerve and muscle in connection with each other through the metal instrument, a very curious phenomenon is observed, the muscles of the almost severed leg becoming spasmodically contracted and then relaxed whenever the contacts were made and broken. Galvani noted the phenomenon first in connection with an electric machine, and looked for an explanation of it in electricity, thinking that there was an analogy between it and the discharge of the Leyden jar. After several years of careful observation, he published a monograph on the subject, which at once attracted attention all over Europe.

Volta was very much interested in Galvani's work, and took up the development of it from the physical side. At first he agreed with the explanation offered by Galvani, who considered that his experiment demonstrated the presence of electricity in animal bodies, and who proposed to introduce the term "animal electricity." After careful investigation, however, Galvani's assertion that animal electricity existed in a form entirely independent of any external electricity, though it had been accepted by most of the distinguished men of science of the time, seemed to Volta without experimental verification. For many years his most determined efforts were used to demonstrate that the muscle twitchings observed were not due to the presence of animal electricity (galvanism as it had come to be called), but to the fact that the metals touching the different portions of the moist nerve muscle preparation really set up minute currents of ordinary electricity.

Some of the experiments which he devised for this purpose were extremely ingenious, and show how thoroughly empirical were his methods and how modern his scientific spirit. In the course of his experiments he found that a difference in the metals of which the arc was composed, when used for the purpose of eliciting the so-called animal electricity, made a great difference in the electrical phenomena observed and in the amount of muscle twitchings obtained. In one brilliant series of experiments, moreover, he showed that, even when the metallic portions touching nerve and muscle were identical, there might still be distinct electrical phenomena, if only an artificial difference in temperature of the end of the metallic arc were produced. Volta was even able to demonstrate that such minute physical differences as the filing of one end of the metallic arc used might give rise to small currents of electricity.

In the midst of these experiments, he came to the realization that two portions of metal of different kinds, separated by a moist non-conducting material, might be made to produce a constant current of electricity for some time. More than this, however, he found that discs of metal of different kinds might be piled on top of one another with intervening discs of moist cloth, and so produce proportionately stronger currents as more and more of the metal plates were employed. This was the origin of the voltaic pile, as it has been called—the first battery for the production at will of regular currents of electricity of definite strength.

While engaged at this he succeeded in demonstrating what has come to be known as Volta's basic experiment; namely, that two plates of metal of different kinds become electrically excited merely by contact. This was practically the beginning of the great advance in applied electricity which ushered in our modern electrical era. It seems a simple matter now, looking back over the century that has elapsed since then, to have taken the successive steps that Volta did for the construction of his electrical pile and for the demonstration of the principle of contact electricity. Groping, as he was, in the dark, however, it took him three years to make the progress that we have described in a few words. How great his discoveries appeared, even to the most distinguished of his scientific contemporaries, can best be judged from an expression of one of the greatest of French electrical scientists, Arago, who declared "Volta's pile the most wonderful instrument that has ever come from the hand of man, not excluding even the telescope or the steam-engine."

An excellent description of just how Volta made his electric pile and what he was able to accomplish with it experimentally in the laboratory, is to be found in the numbers for January and February, 1900, of the Stimmen aus Maria-Laach—a literary and scientific periodical published by the German Jesuits. This article on Alessandro Volta, by Father Kneller, S. J., was written shortly after the celebration of the hundredth anniversary of Volta's invention of the electric pile, when there had just been a fresh sorting over of Volta's documents, and contains a very full set of references to the biographical material for Volta's life. Father Kneller says:

"Before this, no one thought for a moment of any possibility of the practical application of electricity. But all at once the whole situation changed. After eight years of observation and experiment, Volta accomplished one day, at the beginning of 1800, in his laboratory at Como, the construction of an instrument which was to revolutionize the study and the practical applications of electricity. He made a pile composed of a large number of equal-sized copper and zinc discs. On each copper disc he placed one of zinc, and then on this a moistened piece of cloth, and continued the series of alternate discs and cloths in this order until he had a rather high column. This was an apparatus as simple as possible and from which no one but Volta could possibly have promised any results. The inventor, however, knew what he was about.

"As soon as he had connected the upper and lower metal plates by means of a wire, there began to flow from the zinc to the copper a secret something, which by the application of the ends of the wire to muscles caused them to twitch; which appeared before the eye as light; applied to the tongue, gave a sensation of taste; caused a thin wire to glow and even to burn between carbon points; produced a blinding light; decomposed water into its constituents; dissolved hitherto unknown metals out of salts and earth; made iron magnetic; directed the magnetic needle out of its path; inclosed wire coils caused new electric currents to be set up; to say nothing of the awful spectacle which occurred when, under the influence of the electric current, the bodies of executed criminals again gave movements of the limbs, their thoraxes really heaved and sank as if they really breathed, and even a dead grasshopper was caused to spring and apparently to sing again.

"Only now, after the discovery of this new kind of electricity—which did not work merely by jerks, but flowed in a constant stream from pole to pole—only now was this mighty natural agent won to the service of man. Volta is, therefore, above all others, the one who broke ground not only for an immense amount of new knowledge in physics, chemistry and physiology, but who also made possible rapid progress in practical electricity, in telegraphy, in electric motors and power machines, in electroplating and the marvelous results in electro-galvanism which constitute our most wonderful mechanical effects at the present time."

Soon after Volta's discovery of the electric pile, or voltaic pile, as it was called in his honor, his reputation spread throughout Europe. At the beginning of 1800, he sent a detailed description of the voltaic pile to the Royal Society of London. During the year 1801 the scientific journals all over Europe were filled with discussions of his discovery.

The French Academy of Sciences invited him to Paris in order to demonstrate his discoveries to the members of that body. Volta was now looking forward to some peaceful years of study, and, so far as he was personally concerned, would surely have refused the invitation. Circumstances were such, however, that it became a civic duty for him to proceed to Paris.

At this time Napoleon was First Consul, and the Italian cities wished to propitiate his favor as far as possible. It was considered a wise thing by the city to send a special delegation to Paris, and, as they knew Napoleon was deeply interested in scientific discoveries that promised practical results, the name of Volta was suggested as one of the official delegates. As an associate, Professor Brugnatelli, who had made some important investigations in chemistry, and who was later to be an extender of the practical application of Volta's discoveries by the invention of the first method of electroplating, was the other member of the delegation. It is a curious reflection on the facilities for travel at the time, that it took twenty-six days for the delegates to reach Paris from Pavia.

Shortly after their arrival in Paris, the travelers were formally introduced to the members of the French Institute, and a number of sessions of the Academy were held, at which Volta's discoveries were discussed. Volta read a communication on the identity of electricity and galvanism. Napoleon, as First Consul, was present at these sessions in the robe of an Academician, and was not only an interested listener, but occasionally, by pertinent questions, drew out significant details of former experiments and Volta's own theories with regard to the nature of the phenomena observed. At the end of the first meeting, at which Volta took a prominent part, Napoleon spent several hours with him talking about the prospects of electricity.

In his letters to his brothers and to his wife at this time, Volta expressed his pleasure at finding how much attention his discoveries were attracting all over Europe. As he said himself, Germany, France and England were full of them, and all the distinguished scientists were eager to do him honor. In France, he was chosen one of the eight foreign members of the Institute, and was made Knight Commander of the Legion of Honor and of the Order of the Iron Crown. Napoleon selected him as one of the first members of the Italian Academy, which he was then in course of establishing, and conferred on him the honor of Senator and Count of the Kingdom of Italy. The French Academy, after having heard Volta's own description of his experiments and discoveries, contrary to its usual custom, voted to him by acclamation its gold medal. More important still, Bonaparte made him a present of 6000 lire, and conferred upon him an annual income of 3000 lire from the public purse. It is an index of Volta's feeling as a faithful son of the Church, that as this income was allotted to him from the revenues of the bishopric of Adria, he would consent to receive it only after Napoleon's decree had been confirmed by the Pope.

Volta had been for nearly twenty years in the University of Pavia before he finally found for himself a wife. He was then past forty-nine years of age. His wife was the youngest daughter of Count Ludovico Peregrini. She had six sisters, one of whom became a nun, and all the others were married before Volta sought the hand of the youngest. Writing to a friend, he says, "that her sisters had distinguished themselves so much by piety, prudence, good sense and practical economy in their households as well as by the most admirable qualities of heart and mind, that he considered himself very fortunate in obtaining a branch from the family tree; and he took her in preference to others that had been offered to him, even though they were possessed of greater physical beauty, more exalted piety and a larger dowry." The marriage seems to have been a very happy one, notwithstanding the considerable disparity of ages and the very matter-of-fact spirit with which it was entered into by one of the parties at least.

The charming intimacy of his domestic life may be judged from some of his letters to his wife when he was traveling. She was always his confidante with regard to new things in science that he saw, and especially as regards the kindly reception which he met with from scientists and the readiness with which they accepted his views. At first, so many of his ideas were new, that it is not surprising that they were looked at somewhat askance by contemporary scientists. When, on his journeys through France, he noticed the trend of opinion setting in favor of his views in electricity, he took pains to tell his wife, and apparently found his greatest pleasure in having her share the joy of his triumph.

One of the severest blows that he suffered was the untimely death of his eldest son, Flaminio, in 1814. "This loss," he wrote to one of his nephews not long after, "strikes me so much to heart that I do not think I shall ever have another happy day." The relations between himself and his children were all of the kindliest nature; and the character of the man comes out perhaps even more clearly in the traditions that are still extant with regard to the devotion of his servants to him, and especially his body-servant, Polonio. Volta was always a simple and unpretentious person, notwithstanding the fact that scientific and even political honors had been heaped upon him toward the end of his life. It was rather difficult, for instance, to get him to change his old clothes for new ones. This feat was usually accomplished by Polonio, who, when he thought the time had arrived for his master to put on the newer clothes, would engage him in some scientific explanation of a morning; then handing him the new garments, Volta would put them on, and would be wearing them for some time before he noticed it. The old servant was then generally able to persuade him that it was time to make the change. Toward the end of his career, Volta led a retired life in a country house not far from his native city of Como. Foreigners often came to see or even have the privilege of a few words with the distinguished scientist who was regarded as the patriarch of electrical science. To Volta, the being on exhibition was always an unpleasant function. He did not care to be lionized, and frequently refused to allow himself even to be seen unless his visitors had a scientific motive. On such occasions, the only chance of the visitors was to secure the good will of Polonio. He would engage his unsuspecting master in a discussion of clouds or wind, or some appearance in the heavens, or something in the leaves of the neighboring trees, and would then bring him to the portico, that he might see the supposed phenomenon. This would give occasion for the visitors to get at least a glimpse of the scientist, who usually failed to suspect the real purpose for which he had been tempted out of doors.

While thus living in the country, Volta's piety became a sort of proverb among the country people. Every morning at an early hour, in company with his servant, he could be seen with bowed head making his way to the church. Here he heard mass, and usually the office of the day, in which all the canons of the cathedral took part. He had a special place on the epistle side of the altar, not far from the organ. His favorite method of prayer was the rosary. He was not infrequently held up to the people by the parish priest as a model of devotion. Whenever he was in the country, every evening saw him taking his walk towards the church. On these occasions, he was usually accompanied by members of his family, and they entered the church for an evening visit to the blessed sacrament.

His behavior toward those who lived in the vicinity of his country place endeared him to all the peasantry. He was not only liberal in giving alms, but made it a point to visit frequently the houses of the poor and help them as much as possible by counsel and suggestion. His scientific knowledge was at command for their benefit, and he was often able to tell them how to avoid many dangers. He gave them definite ideas with regard to the importance of cleanliness and the necessity of cooking their food very carefully so as to prevent diseases occasioned by badly cooked material. He also taught them to distinguish between the wholesome and the spurred rye, from which their polenta was prepared, in order to escape the dreaded pellagra, the disease so common in Italy, which comes from the use of diseased grain.

He endeared himself so much to the people of the countryside that they invented a special name for him, which proclaimed the tenderness of their liking for the man. They knew how much he was honored for his wonderful discoveries in electricity, and many of them had even seen some of the (to them at least) inexplicable phenomena that he could produce at will by means of various electrical contrivances. At first they called him a "magician"; but as this word has, particularly for the Italian peasantry, a suspicion of evil in it, they added the adjective "beneficent," and he was generally known as *Il mago benefico*.

His interest in these gentle, kindly people may be appreciated from the fact that he knew practically all of his country neighbors by name, and, as a rule, he was familiar also with the conditions of their families and their household affairs. Not infrequently he would stop and talk to them about such things, and this favor was always considered as a precious mark of his neighborly courtesy by the peasantry.

Such was the simplicity of the man whose name is undoubtedly one of the greatest in the history of science. The great beginnings of the chapter on applied electricity are all his. There was nothing he touched in his work that he did not illuminate. His was typically the mind of the genius, ever reaching out beyond the boundaries of the known—an abundant source of leading and light for others. Far from being a doubter in matters religious, his scientific greatness seemed only to make him readier to submit to what are sometimes spoken of as the shackles of faith, though to him belief appealed as a completion of knowledge of things beyond the domain of sense or the ordinary powers of intellectual acquisition. Like Pasteur, a century later, the more he knew, the more ready was he to believe and the more satisfying he found his faith. This is a very different picture of the great scientific mind from that ordinarily presented as characteristic of scientific thinkers. But Volta is not an exception; rather does he represent the rule, so far as the very great scientists are concerned; for it is only the second-rate minds, those destined to follow but not to lead, in science, who have so constantly proclaimed the opposition of science to faith.

Volta's well-known confession of faith declares his state of mind with regard to religion better than any words of a biographer, and it is a striking commentary on the impression that has in some inexplicable way gained wide acceptance, that a man cannot be a great scientist and a firm believer in religion. A distinguished professor of psychology at one of the large American universities said not long since, that a scientist must keep his science and religion apart, or there will be serious consequences for his religion. Volta's opinion in this matter is worth remembering. Having heard it said that, though he continued to practice his religion, this was more because he did not want to offend friends, that he did not care to scandalize his neighbors, and did not want the poor folk around him to be led by his example into giving up what he knew to be their most fruitful source of consolation in the trials of life, while in the full exercise of his intellectual faculties, Volta deliberately wrote out his confession of faith so that all the world of his own and the after time might know it.

"If some of my faults and negligences may have by chance given occasion to some one to suspect me of infidelity, I am ready, as some reparation for this and for any other good purpose, to declare to such a one and to every other person and on every occasion and under all circumstances that I have always held, and hold now, the Holy Catholic Religion as the only true and infallible one, thanking without end the good God for having gifted me with such a faith, in which I firmly propose to live and die, in the lively hope of attaining eternal life. I recognize my faith as a gift of God, a supernatural faith. I have not, on this account, however, neglected to use all human means that could confirm me more and more in it and that might drive away any doubt which could arise to tempt me in matters of faith. I have studied my faith with attention as to its foundations, reading for this purpose books of apologetics as well as those written with a contrary purpose, and trying to appreciate the arguments pro and contra. I have tried to realize from what sources spring the strongest arguments which render faith most credible to natural reason and such as cannot fail to make every well-balanced mind which has not been perverted by vice or passion embrace it and love it. May this protest of mine, which I have deliberately drawn up and which I leave to posterity, subscribed with my own hand and which shows to all and everyone that I do not blush at the Gospel—may it, as I have said, produce some good fruit.—Signed at Milan, Jan. 6th, 1815, Alessandro Volta."

When Volta wrote this, he was just approaching his sixtieth year and was in the full maturity of his powers. He lived for twelve years after this, looked up to as one of the great thinkers of Europe and as one of the most important men of Italy of this time. Far from being in his dotage, then, he was at the moment surely, if ever, in the best position to know his own mind with regard to his faith and his relations to the Creator.

There is a famous picture of Volta, by Magaud, in Marseilles. It chronicles the fact that Volta had become a Count, a Senator and a Member of the French Institute, so appointed by Napoleon, and that he is in some sense therefore a Frenchman. Magaud has painted him standing, with his electric apparatus on one side and the Scriptures on the other. Near him is placed his friend Sylvio Pellico, whose little book, "My Ten Years' Imprisonment," has endeared him to thousands of readers all over the world. Pellico had doubted the presence of Providence in the world and the existence of a hereafter. In the midst of his doubts, he turned to Volta. "In thy old age, O Volta!" said Pellico, "the hand of Providence placed in thy pathway a young man gone astray. Oh! thou, said I to the ancient seer, who hast plunged deeper than others into the secrets of the Creator, teach me the road that will lead me to the light." And the old man made answer: "I too have doubted, but I have sought. The great scandal of my youth was to behold the teachers of those days lay hold of science to combat religion. For me to-day I see only God everywhere."

CHAPTER VI. COULOMB.

Great discoverers in science must usually be satisfied with having their names attached to some one phase of scientific development, be it an instrument, a law, a unit of measurement, a process of investigation or some phenomenon which they first observed. The originality of Coulomb's genius will be better appreciated, since besides having a unit of electricity named after him, there is also a law in electro-magnetics and a torsion-balance that will always be associated with his name. Few men have been more ingenious in their ability to put complex ideas into practical shape and give them simple mechanical expression by instrumental methods. While his name is to be forever associated with the science of electrostatics, he was profoundly interested in other departments of physics, and for him to be interested always meant that he would illuminate previous knowledge by practical hints and suggestions and carry the conclusions of his predecessors a little farther into science than they had ever gone before. His was typically an experimental genius, and he must be considered one of the men of whom not more than half a dozen are born in a century, who are, in Kipling's strong term, "masterless"; who do not need to be taught, but who find for themselves a path into the domain of the unknown.

Coulomb investigated the fundamental law in electricity and magnetism, that attractions and repulsions are inversely as the square of the distances, and showed that it held accurately for point-charges and point-poles. He demonstrated that these interesting phenomena were not chance manifestations of irregular forces, but that they represented a definite mode of action of force, thus setting this department of knowledge on a scientific basis. While in practical significance Ohm's Law, discovered nearly a half century later, is of much more import, Coulomb's discoveries are fundamental in character and, coming in the very beginnings of modern electrical science, did much to guide the infant science in the ways it should follow. The establishing of this law contributed very largely to the rapid development of the twin sciences of electricity and magnetism. It is experimental observation that means most for a rising science;

and, in fact, that Coulomb should have been the pioneer in it stamps him as possessed not only of great originality, but also of the power of independent thinking, which is perhaps the most precious quality for the man of science.

The French investigator succeeded in demonstrating his law by two distinct methods which are still used for illustration purposes in our physical laboratories. In the first, he employed the torsion-balance devised by Michell, and re-invented by himself, an instrument of exact measurement which, in his hands, yielded as invaluable results as it did in those of Faraday half a century later. The instrument depends on the principle first established by Coulomb himself, that when a wire is twisted, the angle of torsion is directly proportional to the force of torsion. In the application of this principle, a fine wire is suspended in a glass case, on the sides of which there is a graduated scale to measure the degree of repulsion between two like poles of a magnet or between similarly electrified bodies.

In his second research on the law of the inverse square, Coulomb used what is known as the method of oscillations. A magnetic needle swinging under the influence of the earth's magnetism is known to act like a pendulum, and as such obeys the laws of pendular motion. In applying this method, Coulomb caused the magnetic needle to oscillate, first, under the influence of the earth's magnetism alone and then under the combined influence of the earth and the magnet placed at varying distances from the needle. The most interesting feature of this work is the manner in which Coulomb succeeded in eliminating the important factor of the earth's magnetism from the problem. It is so simple and ingenious that it commands the admiration of investigators, who employ it in their laboratory work even to the present day.

It is clear, then, that the International Committee which selected the term coulomb for the electromagnetic unit of electrical quantity gave honor where it was eminently due. Coulomb stands out as a man of precision and accuracy, whose methods of exact measurement revolutionized the rising science, and whose researches and discoveries in physics and mechanics furnish ample justification for giving him a place among the makers of electricity. He was one of the gifted men whose original works ushered in so gloriously the nineteenth century, and who laid the deep and firm foundations on which the last three generations have built up the magnificent temple of electrical science.

Charles Augustin de Coulomb was born at Angoulême, June 14th, 1736. His ancestors for several generations had been magistrates, and were looked upon as representatives of the country nobility. He made his university studies in Paris, and while still young, entered the army. From the very beginning, however, his genius for mathematics was recognized, and he was employed in the capacity of military engineer. To Americans, it will be interesting to know that his first engineering project was undertaken at Martinique, where he constructed Fort Bourbon. His sterling character and remarkable ability secured him rapid advancement in the service. In spite of the fact that the climate did not agree with him, he remained for three years on the island, because he would not employ the political influence that might have secured his recall, since he thought it his duty to serve his country in an important colonial post. Nearly all his comrades perished by fever. It is the irony of fate that after his return to France a change in the ministry deprived him of the just recompense of his devotion to country, and he did not receive the special extraordinary promotion which he had earned in this special detail.

During a short stay that he made at Paris after his return, he sought the society of men of science as far as possible, and succeeded in getting in touch with all that was most promising in scientific progress at the time. He was already known rather favorably by many of the scientific men of the capital because of the paper on The Statics of Vaults, a monograph on static problems in architecture, which he presented to the Academy of Sciences in 1779. His next military assignment was to Rochefort. Here he composed his monograph on "The Theory of Simple Machines," which carried off the double prize that had been offered by the Academy of Sciences for the solution of problems connected with this important question. This attracted the attention not only of the scientific world, but also of his military superiors. As a result, he was sent successively to Cherburg and to the Isle of Aix, to direct engineering works, and accomplished the tasks involved with success.

Two years later, when he was about forty-five years of age, he was elected member of the Academy of Sciences by a unanimous vote. He was a man of great personal magnetism, and all those who came in contact with him learned to like him for his straightforward character and for the absolute righteousness of his life. Few men have made firmer friends than Coulomb, as few have ever shown more unselfish devotion to duty and to conscience than he, though under circumstances that were neither spectacular nor theatrical. It was harder to face the deadly climate of Martinique than it would have been to take one's place at the head of a forlorn hope in an outburst of enthusiastic courage; and Coulomb was to have other trials of quite as deterrent a nature, and was to meet them with the same imperturbed sense of duty.

Graft is sometimes supposed to be temptation peculiar only to our own times, but the opportunities for it have always been present in such work as Coulomb had to oversee, and the army engineer of all ages has had to stand or fall before it. It was proposed, about this time, to build a system of government canals in Brittany. Such a canal-system would, as is easy to understand, cost an enormous sum of money and give magnificent opportunities for speculation of various kinds. No small objection had been made to the project, on the score that it would not confer all the benefits on the region that were claimed for it, and Coulomb was commissioned by the Minister of Marine to determine the question of the advisability of constructing the canals, and of the probable effect which they would have on the commerce of the country.

After careful investigation, he came to the conclusion that the advantages which were expected to accrue from the project would not compensate for the enormous expense that would be entailed.

This decision aroused the angry protest of a strong political faction, who expected to reap wealth and personal advantages of many kinds from the scheme, and who protested bitterly against Coulomb's report. He was able to support his conclusions in the matter, however, with such unanswerable mathematical and engineering arguments, that his opinion prevailed and the project was given up.

As a consequence, instead of the opportunity to serve a political party with every avenue to preferment and, above all, to wealth open for him, he found himself, for the time being, deprived even of the opportunity to devote himself further to his favorite occupations in military engineering. The excuse given for this interruption in his career, for there has always been an excuse for such action, was that proper representations for permission to make the report had not been made to the Minister of Marine; and instead of commendation, Coulomb received what was practically a reprimand.

Wounded by this injustice, which was manifestly due to the fact that his honest report had displeased those who expected to reap personal benefit from the canal project, and disgusted with a service in which such things were possible, Coulomb sent in his resignation. The Minister of Marine realized that the acceptance of the proffered resignation would surely expose the ministry at least to suspicion as to the reasons why Coulomb's report was not accepted with good grace. Permission to retire from the service was refused, as this would insure his silence. He was ordered back to Brittany to continue his work there, possibly with the idea that this unfavorable experience would be sufficient of itself to make him understand what was expected of him and render him a little more complacent to the wishes of those in authority. If any such ideas were entertained, they were destined to grievous disappointment. Coulomb was not of those who, seeing duty plainly, refuse to follow it because some personal advantage or disadvantage intervenes. Selfish reasons did not appeal to his character nor obscure the issues.

He went back to Brittany, ready to express his firm opinion in the matter and with integrity of soul untouched. The consequence was that the provincial authorities, recognizing their true interests, acknowledged the error they had come near falling into, and now wished to reward the engineer handsomely for his unswerving devotion to duty. Coulomb as promptly refused a reward for doing his duty as he had ignored even the appearance of a bribe to avoid it. Only after considerable pressure was he prevailed upon to accept the best timepiece they could procure, on which the arms of the province were engraved. It had what was quite rare in those days, a second's hand, and he constantly made use of this in all his experimental work thereafter. A French biographer says that, never was a souvenir better chosen nor more suitably employed. Coulomb's merits were recognized by the government authorities not long after, and he was made superintendent of the fountains of France. A few years later, he was promoted to the position of Curator of Plans and Relief Maps of the Military Staff of France, and was chosen as one of the commission of the French Academy of Sciences who went to England in order to study hospital conditions there. At this time, he was at the acme of his career. His grade was that of Lieutenant Colonel of Engineers, a position much higher in the foreign armies at that time than would be the post with the corresponding title in our army. He had been made a Chevalier of St. Louis, and it looked as though a brilliant future were opening out before him. Each year, for a decade, had seen the publication of one or more memoirs on important subjects, nearly every one of which contained some original material of the highest value, destined not only to add to Coulomb's reputation, but to furnish basic information for the further development of science.

In 1789, however, the Revolution broke out, and there was an end to all Coulomb's opportunities for work. He was utterly out of sympathy with the movement, the worst consequences of which he foresaw from the beginning, and he at once handed in his resignation of the various positions that he occupied under the government. He went into almost absolute retirement, devoting himself to the education of his children. During this time, however, he did not cease to cultivate science, inasmuch as he gave the finishing touch to various papers which he had previously outlined. Unfortunately, however, his departure from Paris made it impossible for him to continue his investigations in electricity for want of apparatus, and so there is a ten years' interruption in his life of scientific activity and of original work. Besides, it cannot be surprising that he should not have had the heart to go on with his work under the awful social conditions that prevailed. Many of his friends lost their lives during the stormy period of the Revolution; most of the others were banished or were in hiding. His beloved country had gone into an unfortunate eclipse, as he could not help but consider it; most of the nations of the earth were indeed in league against her, and the end was not yet in sight. It would be too much to expect of human nature that it should devote itself to abstruse problems in science at moments of such disturbance as this, and so some of the possibilities of Coulomb's original genius were lost to science during that calamitous period.

Like many of the great discoveries of science, Coulomb's most important work was done in the course of other investigations, and came by what might be called a happy accident. He had been investigating the qualities of wire of various kinds, especially with regard to their elasticity, so as to be able to determine the limits of their use in various engineering projects. When he discovered that the elasticity of torsion of a wire was a constant property, he proceeded to utilize it in the calculation of such delicate phenomena as those of electric and magnetic forces. The first instrument for this purpose that he constructed consisted simply of a long magnetized needle suspended horizontally by a fine wire. Supposing this needle to be at rest, if one moves it away from the magnetic meridian by a certain number of degrees, the twisted wire will have a definite tendency to untwist and to bring back the needle to its original position by a series of oscillations whose frequency can be readily observed.

For such observations, it is possible to obtain the value of the force acting on the needle and causing it to move to and fro at a given rate. This was the underlying idea which received very simple expression in the ingenious instrument which Coulomb devised and called a torsion-balance. With it, he set about determining the law which governs the mutual action of magnets and of electrified bodies with regard to distance, and found it to be the same as that which Newton found to hold for bodies distributed throughout the universe, that is, that attraction and repulsion vary inversely as the square of the distance. He also proved, with the aid of his torsion-balance, that the forces of attraction and repulsion vary as the product of the strength of the poles in one case and as the product of the electric charges in the other. These were the important discoveries of Coulomb's life; they served to earn for him the right to have his name given to the unit of electrical quantity, the *coulomb*.

Coulomb did not stop here, however, but proceeded to apply his laws to various other phenomena. He proved that electricity distributes itself entirely over the surface of a body without penetrating the mass of the conductor, and he showed by calculation that this result was a necessary consequence of the law of repulsion.

A list of the papers which he published on electricity and magnetism, the titles of which, with French accuracy of expression, furnish an excellent idea of their contents, shows the thoroughly progressive and scientific spirit of the man, and how well he proceeded from the known to the less known, always widening the bounds of knowledge. Suffice it to say here that the observations of Coulomb were not only original, but that they concerned some of the most difficult questions in electricity, and that he was clearing the ground for others in such a way as to make future work and quantitative measurements in electricity reliable and comparatively easy. It is because of this pioneer work that Coulomb deserves so much praise. It was not long before Coulomb's observations were confirmed by others, and then the beginnings of the modern development of electricity became manifest, owing not a little to the researches and inventions, the genius and ingenuity of this French military engineer.

Some phases of electrical development attributed to others really belong to Coulomb. A typical example of this detraction from his merit is the attribution to Biot of the solution of the problem of the complete discharge of an electrified sphere by means of two hollow hemispheres. This experiment is fully described by Coulomb, and he even emphasizes the fact that the external discharging bodies need not necessarily be of the same shape as the charged sphere. Some of what Coulomb accepted as principles in electricity have proved in the course of time, not to be the realities that he thought them; but the progress that has led to such contradictions of his opinions has been mainly rendered possible by his own discoveries. The fable of the eagle stricken by the arrow containing some of its own feathers, is so old that one might think that, when the progress of a science due to a scientist brings men beyond the position he occupied, they would not blame him for backwardness. This is, however, one of the curious critical methods in the history of science that has most frequently to be deprecated by the historian who is tracing origins and developments.

Coulomb's papers, with the exception of his memoir on "Problems in Statics Applied to Architecture," his "Researches on the Methods of executing Works under Water without the Necessity of Pumping," his "Theory of Simple Machines," and his researches "On Windmills," which form separate monographs, were all published together in a single volume by the French Physical Society in 1884.

This volume contains, besides his investigations on the best way of making magnetic needles, his theoretic and experimental investigations on the force of torsion and on the elasticity of metallic threads, which were undertaken in order to enable him to make his electric torsion-balance something more than mere guess-work. All the other papers are concerned directly with electricity or magnetism, and show how actively, nearly a hundred and twenty-five years ago, a great mind was engaged with problems in electricity which we are apt to consider as belonging more properly to our own time. The list of papers published in these memoirs, arranged in chronological order, gives a good idea of the development of electrical science in Coulomb's own mind. There is a logical as well as a chronological order to be observed in them.

In 1785, when he was just approaching his fiftieth year, there were three subjects with regard to which Coulomb's experimental observations enabled him to set down some definite principles. The first of these was the construction and use of an electric balance, founded on the property which wires have of exhibiting a torque proportional to the angle of torsion. The second was the determination of the laws, according to which the magnetic and electric "fluids," as Coulomb and investigators in electricity called them at that time, act both as regards repulsion and attraction. The third was the determination of the quantity of electricity which an insulated body loses in a given time from contact with air more or less moist.

In 1786, he published a paper in which he demonstrated what he considered the principal properties of the electric fluid. These are, that this fluid does not spread itself on a substance by any chemical affinity or any elective attraction, but that it distributes itself over various bodies that are placed in contact, entirely in accordance with their shape; and also that in electrical conductors, the charge is limited to the surface of the conductor and does not penetrate to any appreciable depth.

In 1787, his only paper was on the manner in which the electrical fluid divides itself between two conducting bodies placed in contact, and on the distribution of this fluid over the different parts of the surface of these bodies. He continued his investigations into this subject in 1788, and also succeeded in determining the density of the electricity at different points on the surface of conducting bodies.

In 1789, he began to work more particularly on magnetism. His first paper on the subject was published that year. Unfortunately, as we have said, the Revolution interrupted his scientific investigations at this point, and for the next eleven years we have nothing from his pen. As a nobleman, he was compelled to leave Paris, and this not only put him out of touch with scientific work generally, but deprived him of the opportunities of using such apparatus as was necessary to carry on his experiments. That he acted prudently in leaving Paris, the careers of other scientists amply prove. Lavoisier continued to carry on his chemical investigations during the stormy times of the Revolution, but his stay in the capital eventually cost him his life. Abbé Haüy, the father of crystallography, who, because of his contributions to the science of pyro-electricity, is of special interest to us, continued to work at his crystals throughout even the Reign of Terror. When thrown into prison, he asked and obtained permission to have his crystals with him. His friends saved him from Lavoisier's fate, but not without an effort, as his life was seriously endangered.

It is easy to understand, however, that a member of the nobility like Coulomb, whose life had been spent in military affairs, should not be able to devote himself seriously to scientific matters while his country was in such a turmoil.

In 1801, he resumed his investigations once more, but now they are concerned more particularly with magnetism. The first was a theoretical and practical determination of the forces which hold different magnetic needles, magnetized to saturation, in the magnetic meridian. This was followed, in the same year, by a paper which, like its predecessor, was published among the memoirs of the Institute of France, which had replaced the Royal Academy of Sciences, to which body many of Coulomb's papers of the former time had been presented, and in whose publications they originally appeared. This second paper detailed his experiments on the determination of the force of cohesion of fluids and the law of resistance in them, when the movements were very slow.

When the French Institute was organized under Napoleon in 1801, Coulomb was named among its first members. It is believed that he was even chosen to occupy a place in the first government of the state, but a man more interested in politics obtained the place, a fortunate circumstance for science. Coulomb was named, however, one of the inspectors of public instruction, then the highest place in the education department, and he did much to restore to France the educational system that had been destroyed during the Revolution. In this rather trying work he was noted for the kindliness yet firmness of his character, while his absolute fairness and sense of justice were recognized on all sides.

Unfortunately Coulomb was not long spared to continue his work. He took up his experimental and mathematical investigations, on his return to the capital, with great enthusiasm, but his health had been undermined and his work had been rudely interrupted. After 1801, no further paper by him appears to have been published until 1806. This gave the result of different methods employed in order to produce in blades and bars of steel the greatest degree of magnetism. For some time preceding this, in spite of increasing ill-health, he had continued his experiments on the influence of temperature on the magnetism of steel. His work on this subject was not destined to be completed, for not long after passing his seventieth year, in June of this year, his health gave way completely, and he died August 23d, 1806. His final observations were gathered by Biot, carefully preserved, and assigned a place in the volume of Coulomb's Memoirs, issued by the French Physical Society.

Personally, Coulomb was noted for great seriousness of character, though with this was mingled a gentleness of disposition that made for him some cordial friendships among his scientific contemporaries. He had but few friends, but those who were admitted to his intimacy made up by the depth of their affection for the smallness of their number. Even those who had occasion to meet him but once or twice, carried away from their meeting an affectionate remembrance of his kindliness and courtesy and readiness to help wherever he could be of service. He was extremely happy in his family relations, and this proved to be a great source of consolation to him during the years when the progress of the French Revolution took him away from science and made him almost despair of his country.

It is not surprising that Biot, the great French physicist, in writing of Coulomb in his Mélanges Scientifiques et Littéraires, Vol. III. (Paris, 1858), should have held Coulomb up as a model of the simple, earnest, helpful life and as a man of the most exemplary character.

He says: "Coulomb lived among the men of his time in patience and charity. He was distinguished among them mainly by his separation from their passions and their errors, and he always maintained himself calm, firm and dignified *in se totus teres atque rotundus*, as Horace says, a complete, perfect and well-rounded character." Few men have deserved so noble a eulogy as this, written nearly fifty years after his death, by one who had known Coulomb himself and his contemporaries well; it has none of the exaggeration of a funeral panegyric, and is evidently founded on details of knowledge with regard to the great electrician which had become a tradition among French scientists, and which Biot has forever crystallized into the history of science by his emphatic expression.

One could scarcely wish for a better epitaph than Biot's summing up of Coulomb's personal character: "All those who knew Coulomb know how the gravity of his character was tempered by the sweetness of his disposition, and those who had the happiness to meet him at their entrance into a scientific career have kept the most tender remembrance of his gentle good-heartedness."

CHAPTER VII. HANS CHRISTIAN OERSTED.

Whatever may be thought of the value of controversy in other departments of knowledge, it has certainly proved useful in the progress of experimental science. Witness the animated and prolonged discussion which took place between Volta and Galvani, and which led to enduring results for the welfare of mankind. Wishing to prove the correctness of his theory of electrification by contact against Galvani's animal electricity, Volta devoted himself unremittingly to experimentation until, in the century year 1800, his brilliant work culminated in the invention of the "pile" or electric battery which bears his name.

A suspicion had been growing for many years in the minds of physicists, that there must be some degree of relationship, probably an intimate one, between magnetism and electricity, between magnetic and electric forces. In the year 1785, van Swinden, a celebrated Dutch physicist, published a work on electricity in which he described and commented upon a number of analogies which he had observed between the two orders of phenomena; but, voluminous as was the work, it threw no light on the nature of the suspected relationship.

It was well known, in the case of houses and ships struck by lightning, that knives, forks and other articles made of steel were often found to be permanently magnetized. Following up this pregnant observation, experimenters often sought to impart magnetic properties to steel needles by Leyden-jar discharges, but with indifferent success. Sometimes there would be a trace of magnetism left and sometimes none. In no case was it possible to say beforehand which end of the knitting-needle would have north polarity and which south.

Though we are better equipped to-day for research work than were our predecessors in the electrical field fifty years ago, we are still unable to predict the polarity that will result in a bar of iron from a given condenser discharge. The uncertainty arises from the fact disclosed by Joseph Henry in 1842 and well known to-day that, under ordinary circumstances, all such discharges consist of a rush of electricity to and fro, that is, they give rise to an oscillatory current of exceedingly short duration. Were it otherwise, that is, were the discharge unidirectional, the needle would always be magnetized to a degree of intensity proportional to the energy released; and it would be possible in every case to foretell with certainty the resulting polarity which the needle would acquire.

With the advent of the voltaic battery, a generator which supplies a steady flow of current in one direction, the interesting problem of relationship between electric and magnetic forces was again attacked; and this time with considerable success.

Probably the earliest investigator afield was Romagnosi, an Italian physician residing in Trent (Tyrol), who, in the year 1802, published in the "Gazetta" of his town an account of an experiment which he had made, and which showed that he was working on promising lines. What he did was this: having connected one end of a silver chain to a voltaic pile, and having carried the chain through a glass tube for the purpose of insulation, he presented the free end, terminating in a knob, to a compass-needle, also insulated. At first, the needle was attracted; and, after contact, repelled. Whatever Romagnosi thought of his experiment and its theoretical bearing, the attraction and subsequent repulsion of the compass-needle which he said he observed were electrostatic and not electromagnetic effects. The Italian physician was indeed on the verge of a great discovery; but he halted in his course and lost his opportunity.

Mojon, Professor of chemistry in Genoa, was a little more fortunate, though he, too, failed to improve his opportunities. In 1804, he sought to magnetize steel needles by placing them for a period of twenty days in circuit with a battery of one hundred elements of the crown-of-cups type, and had the satisfaction of finding them permanently magnetized when withdrawn from the circuit. Unlike the electrostatic effect of his fellow-countryman Romagnosi, this was unquestionably an electromagnetic effect, the first link in the long chain connecting electricity with magnetism.

That this result attracted wide attention at the time, as it well deserved, is evident from the notice given by Izarn in his "Manuel du Galvanisme," and by Aldini in his "Essai Théorique et expérimental sur le Galvanisme," both of which were published in Paris in the same year, 1804.

Though the manuals of Izarn and Aldini served to give a fresh impetus to the quest of the relationship between electricity and magnetism, it was not, however, until the year 1820 that the cardinal discovery was made by one philosopher and the intimate relationship revealed by another. Then all Europe rang with the names of Oersted, the fortunate discoverer of the "magnetic effect" of the electric current, and Ampère, whose masterly analysis disclosed the nature of the long-sought-for connection. In the delight of the hour, men called Oersted the Columbus, and Ampère the Newton, of electricity.

Though a philosopher of a high order and lecturer of interest and brilliancy, Oersted was, nevertheless, a poor experimentalist. He was fine in the abstract, awkward in the concrete. Often did he call for the assistance of a student to perform an experiment for the class under his direction. Hansteen, who is celebrated for his very fine work in terrestrial magnetism, often had this privilege, for he was clear of mind and deft of hand. Writing to Faraday, he said: "Oersted was a man of genious, but very unsuccessful as a demonstrator, for he could not manipulate instruments."

In seeking for some evidence of a physical interaction between electricity and magnetism, Oersted on one occasion, placed a wire conveying a current vertically across a compass-needle; and, on obtaining no result, seemed greatly disappointed. He evidently expected the needle to respond in some way to the energy of the current; and so it would have responded had he placed the wire in any other position than the particular one which he selected. The Danish philosopher now hesitates; and for lack of coolness, patience and resourcefulness, runs the risk of losing the crowning glory of his life. He is disappointed at his failure; and for the nonce, contents himself with brooding over it.

On another occasion, having a stronger battery at his disposal, he determined to try the experiment again, in the hope that the greater energy at his command would provoke the magnet to respond. This time, he stretched the wire over and parallel to the compass needle, when, to his intense delight, the magnet turned aside as soon as the circuit was closed. The result was pronounced and instantaneous. The Professor, an enthusiast by nature, waxed warm over his good fortune, and well might he do so, as the discovery which he had just made was destined to revolutionize existing modes of transmitting intelligence to distant parts and bring remotest countries into direct, and immediate relation with one another.

That Oersted fell into ecstasy over his success was but natural, though it is not stated that he exhibited his enthusiasm by the performance of any unusual feat. When Lavoisier made a discovery, he was wont to take hold of his assistant and go dancing around with him for sheer joy. After making a certain successful experiment in his laboratory, Gay-Lussac gave vent to his feelings by dancing round the room, and clapping his hands the while. It is related that, when Davy saw the first globules of potassium burst through the crust of potash and take fire, his delight knew no bounds. He also took to dancing, and some time had to elapse before he was sufficiently composed to continue his work. Even the cool and self-possessed Faraday occasionally waxed warm on seeing his efforts crowned with success. It is said that, when he got a wire conveying a current to revolve round the pole of a magnet, he rubbed his hands vigorously and danced around the table, his face beaming with delight: "There they go, there they go; we have succeeded at last," he said. He then gleefully proposed to cease work for the day and spend the evening at Astley's seeing the feats of well-trained horses!

Having realized that his experiment was one of fundamental importance in physical theory, our philosopher proceeds to repeat it under varying conditions. He places the wire conveying the current in front of the needle, behind it, under it, across it; he reverses the current in each case, and notices the direction in which the needle turns. Though he states results very clearly, he gives no general rule whereby the direction of the deflection may be foretold from that of the current. A *memoria technica* to meet all cases that may occur was needed, and was promptly supplied by Ampère, who, with a flash of genius, devised the rule of the little swimmer. Others have been added since, such as the cork-screw rule and the rule involving the outspread right hand; but the swimmer appeals in a manner quite its own to the fancy of the youthful student. It pleases while it instructs; it is ingenious while yet remarkably simple.

It has been said that the Philosopher of Copenhagen was led by mere accident to the experiment which will hand his name down the ages; but inasmuch as he was looking, during thirteen years, for a result analogous to the one which he obtained, it is only right to give him full credit for the success which he achieved. It has been well remarked, that the seeds of great discoveries are constantly floating around us, but take root only in minds well prepared to receive them. Accidents of the Oersted type happen only to men who deserve them, as was the case with Musschenbroek and Galvani in the eighteenth century, and with Roentgen in the nineteenth. The electrification of a flask of water, the twitching of frogs' legs in response to electric sparks, and the blackening of a sensitive screen by a distant, shielded Crookes's tube, led to the electrostatic condenser in the first case, to "galvanism" in the second, and to the photography of the invisible in the third.

Writing of Oersted's discovery, Faraday said that "It burst open the gates of a domain in science, dark till then, and filled it with a flood of light."

The discovery of 1820 was hailed throughout Europe by an extraordinary outburst of enthusiasm. Oersted was complimented and congratulated on all sides. Honors were showered upon him: the Royal Society of London awarded him the Copley medal; the French Academy of Sciences gave him its gold medal for the physico-mathematical sciences; Prussia conferred upon him the Ordre pour le Mérite, and his own country made him a Knight of the Daneborg.

Oersted lost no time in preparing a memoir on the subject of his work, a copy of which was sent to the learned societies and most renowned philosophers of Europe. The memoir, which was written in Latin and dated July 21st, 1820, consisted of four quarto pages with the title "Experiments on the effect of the electric conflict on the magnetic needle."

A perusal of this paper brings home the conviction that Oersted realized fairly well the forces which came into play in his experiment; for in one place, he speaks of the effect as due to a transverse force emanating from the conductor conveying the current, and again as a conflict acting in a revolving manner around the wire. A complete statement of the nature of the mechanical force exerted by a conductor conveying a current on a magnetic needle was given almost immediately by Ampère, a master analyst and accomplished experimentalist.

It will stand for all time in the history of science, that in less than two months after the publication of Oersted's memoir, Ampère succeeded in showing the mechanical effect in magnitude and direction of an element of current not only on the magnetic needle itself, but also on a similar element of an adjacent conductor conveying a current, thereby founding a new science in the department of physics, the science of electro-dynamics.

Oersted does not appear to have given thought to the practical possibilities of his discovery. While appreciating the utilitarian in science, he evidently preferred the pursuit of knowledge for its own sake. In a discourse which he delivered in 1814 before the University of Copenhagen, he put himself on record when he said that "The real laborer in the scientific field chooses knowledge as his highest aim."

So said Plato ages before, and so said Archimedes, who held that it was undesirable for a philosopher to seek to apply the discoveries of science to any practical end. The screw which he invented, his catapults and burning mirrors, show, however, that when necessary the Syracusan mathematician could come down from the serene heights of investigation to the prosaic arena of application.

Before Oersted spoke of "the real laborer," Thomas Young had affirmed that "Those who possess the genuine spirit of scientific investigation are content to proceed in their researches without inquiring at every step what they gain by their newly discovered lights, and to what practical purposes they are applicable."

Magnetic Whirl Surroundinga Wire Through Which a Current is Passing

Young's most illustrious successor in the Royal Institution, Michael Faraday, devoted himself calmly but unflinchingly to research work, in the conviction that no discovery, however remote in its nature, from the subject of daily observation, could with reason be declared wholly inapplicable to the benefit of mankind. After discovering in 1831 that electric currents could be produced by the relative motion of magnets and coils of wire, a discovery which is the basis of all the electric engineering of our day, Faraday constructed several experimental machines embodying this principle, and then turned away abruptly from the work, saying, "I had rather been desirous of discovering new facts and new relations dependent on magneto-electric induction than of exalting the force of those already obtained, being assured that the latter would find their full development hereafter."

Our own Joseph Henry, whose sterling merit is universally recognized, beautifully said in this connection: "He who loves truth for its own sake feels that its highest claims are lowered by being continually summoned to the bar of immediate and palpable utility."

Oersted seems to have shared the opinion largely held by the scientific men of his day, that electricity is mainly a magnetic phenomenon. Ampère, for one, did not think so, as is evident from the beautiful theory which he devised to explain the magnetism of a bar by minute electric currents flowing round each individual molecule of the iron. To the French physicist, magnetism was purely an electrical phenomenon.

Though propounded more than eighty years ago, this theory is still in harmony with all facts and phenomena in the domain of magnetism known to-day. It is important to remember, when thinking of this physical theory, that the Amperian currents in question are confined to the molecule, and that they do not flow from one molecule to another. Critics have urged against the theory that the molecules must be heated by the circulation of these elementary currents, to which objection it has been replied that, as we know nothing of the nature of the molecule, we cannot say that it offers any resistance to the current; and, therefore, we cannot affirm that there is any development of heat due to the circulation of these elementary currents.

It is to Ampère's credit that he was also the first to propose a practical application of Oersted's discovery, an application that was nothing less than the electric telegraph itself. He suggested that the deflection of the magnetic needle could be used for the transmission of signals from one place to another by means of as many needles and circuits as there are letters in the alphabet. If Ampère had only recalled the optical and mechanical telegraphs in use in his day, such as the swinging of lanterns by night and wigwagging of flags and the movements of semaphores by day, he might have reduced his twenty-four circuits to one, using the two elements, viz., motion of the needle to the right and motion to the left, to make up the entire alphabet. Morse substituted the dot and the dash for these deflections, and thus rendered the reception of messages automatic and permanent.

In connection with this proposal to use a magnetic needle for the transmission of intelligence, the reader will no doubt recall the lover's telegraph, so beautifully described by Addison in the "Spectator" for December 6th, 1711; but ingeniously conceived as it was, this magnetic telegraph was purely and simply a creation of the imagination.

This canny conceit has been attributed to Cardinal Bembo, the elegant scholar and private secretary to Pope Leo X.; but it was his friend Porta, the versatile philosopher, who made it widely known by the vivid description which he gave of it in his celebrated work on "Natural Magic," published at Naples in 1558.

This sympathetic telegraph consisted, we are told, of a magnetic needle poised in the center of a dial-plate, with the letters of the alphabet written around it. The two fortunate individuals privileged to hold *wireless* correspondence with each other having agreed as to the day and the hour, proceed to the room in which the wonderful instrument is kept, where, as soon as one of them turns the needle of his transmitter to a letter, the distant needle turns at once in sympathy to the same letter on its dial!

Such is the power of magnetic sympathy, that the instruments will work successfully though hills, forests, lakes or mountains intervene! Porta has it: "To a friend at a distance shut up in prison, we may relate our minds; which, I do not doubt, may be done by means of compasses having the alphabet written around them."

This sympathetic magnetic telegraph figures extensively in the scientific literature of the sixteenth and seventeenth centuries: some believed in the figment, others condemned it. Addison described it in elegant prose, and Akenside in beautiful verse. Perhaps the most famous composition on the subject is a short Latin poem, written, after the style and vein of Lucretius, in 1617 by Famianus Strada, an Italian Jesuit. A few years after its publication in the author's "Prolusiones," a metrical translation was made by Hakewill and inserted on page 285 of his "Apologie, or Declaration of the Power and Providence of God," 1630.

Owing to the interest that attaches to this celebrated composition and the difficulty of getting Hakewill's "Apologie," we append his version of the poem.

The Loade above all other stones hath this strange property If sundry steels thereto or needles you apply, Such force and motion thence they draw that they incline To turn them to the Bear, which near the Pole doth shine. Nay, more, as many steels as touch that virtuous stone In strange and wondrous sort conspiring all in one Together move themselves and situate together: As if one of those steels at Rome be stirred, the other The self-same way will stir though they far distant be, And all through Nature's force and secret sympathy; Well then if you of aught would fain advise your friend That dwells far off, to whom no letter you can send; A large smooth round table make, write down the crisscross row In order on the verge thereof, and then bestow The needle in the midst which touch'd the Loade that so What note soe'er you list, it straight may turn unto. Then frame another orb in all respects like this Describe the edge and lay the steel thereon likewise, The steel which from the self-same Magnes motion drew; This orb send with thy friend what time he bids adieu. But on the days agree at first, when you do mean to prove If the steel stir, and to what letter it doth move. This done, if with thy friend thou closely wouldst advise, Who in a country off far distant from thee lies, Take thou the orb and steel which on the orb was set The crisscross on the edge thou seest in order writ. What notes will frame thy words, to them direct thy steel And it sometimes to this, sometimes to that note wheel Turning it round about so often till you find You have compounded all the meaning of your mind. Thy friend that dwells far off, O strange! doth plainly see The steel so stir though it by no man stirréd be, Running now here, now there: he conscious of the plot As the steel-guide pursues, and reads from note to note. Then gathering into words those notes, he clearly sees What's needful to be done, the needle truchman is. Now, when the steel doth cease its motion; if thy friend Think it convenient answer back to send, The same course he may take; and, with his needle write Touching the several notes which so he list indite. Would God, men would be pleased to put this course in use, Their letters would arrive more speedy and more sure, No rivers would them stop nor thieves them intercept; Princes with their own hands, their business might effect. We scribes, from black sea 'scaped, at length with hearty wills At th' altar of the Loade would consecrate our quills.

Another translation of the poem was made by Dr. Samuel Ward and published at the end of his "Wonders of the Loadstone," 1640.

Ampère's suggestion, made, as we have seen, in the year 1820, was not the first proposal to use electricity for telegraphic purposes. Already, in 1753, a writer in *The Scots Magazine*, signing himself C. M. (Charles Morrison, of Greenock, according to Sir David Brewster, and Charles Marshall, of Paisley, according to Latimer Clark), outlined a method involving the use of frictional electricity; and Lesage, of Geneva, constructed a short experimental line, in 1774, consisting of twenty-four wires and a pith-ball electroscope. But the man who attained the greatest success in the employment of static electricity for this purpose was Ronalds, of London, who, in 1816, erected a single-wire line eight miles long in his gardens at Hammersmith, with a pair of pith-balls and a rotating disc for receiving instrument.

When well satisfied that his system was practicable and reliable, Ronalds wrote to the head of the intelligence department in London urging the adoption of his invention for the public service; but he was promptly brought to realize the scant encouragement so often extended to inventors by persons in high places, that responsible official politely informing him "that telegraphs of all kinds are wholly unnecessary," and that no other than the mechanical one in daily use would be adopted.

When penning these words, the representative of the British government must have forgotten the experience of 1812, when the result of the battle of Salamanca was semaphored from Plymouth to London, on which occasion a fog cut off the message after the transmission of the first two words, "Wellington defeated," the remainder of the despatch, "the French at Salamanca," reaching the capital only on the following morning!

A rapid sketch of the life of our philosopher, whose discovery of the magnetic effect of the voltaic current in 1820 led to the invention of the electric telegraph, cannot be without interest.

Hans Christian Oersted was born on August 14th, 1777, in the little town of Rudkjöbing, in the island of Langeland, Denmark. Being the son of poor parents, his early years were spent in very narrow circumstances. He and his younger brother were mainly indebted to their own efforts for whatever instruction they received in the rudiments of learning. The town in which they lived being small, offered few opportunities for education, even if the family exchequer had been such as to permit the boys to take advantage of them. There was a German wigmaker in the place, however, who was a little more advanced in knowledge than the generality of the townspeople. He and his wife liked the Oersted boys, who were very frequently to be found in the wigmaker's shop. The good housewife taught them to read, while the artist himself taught them a little German. Hans Christian advanced so rapidly in his studies that he acquired a reputation for precociousness, which, with the usual prejudice against bright children, made the neighbors shake their heads prophetically and say: "The child will not live; he is too bright to last long."

Hans Christian learned the elements of arithmetic from an old school-book which he picked up by chance; and no sooner had he advanced a little, than he set about instructing his brother. Very probably, the teacher benefited quite as much by this process of instruction as the pupil. Adversity is a good school for the formation of character as well as for the acquisition of knowledge. It is evident, from the lives of such men as Oersted, Faraday, Kepler, Ohm, and others who were brought up in the lap of poverty, that it is not so much educational opportunity that is needed for the development of mind which we call education, as the earnest determination and the abiding desire to have it. Even boyhood creates its own opportunities for education despite intervening obstacles, if it has only a decided eagerness, a pronounced thirst for knowledge.

About the time that the young Oersteds entered their teens, their father secured the services of a private teacher to give them some instruction in the rudiments of Latin and Greek. This accidental preceptor was only a wandering student who happened to be in the place at the time; but the boys, in their eagerness to learn, derived more benefit from his lessons than many boys of their age often do nowadays from the help and encouragement of a carefully selected and academically equipped tutor.

At the age of twelve, Oersted senior was taken into his father's apothecary-shop in quality of assistant, a position which seemed destined to put an end to all opportunities for further advancement in the path of learning. When a boy goes into a drug-store in an official capacity, his future career is usually settled; he is a druggist to the end. His new avocation, however, proved to be the beginning of new intellectual activities for Oersted. The chemical side of his work became a source of new information to him, and also a stimulus to learn all that he could of chemistry and kindred subjects. Science became a hobby with the young apothecary, and everything relating to it appealed to him. What Hans learned, he as usual imparted to his brother, who was already becoming interested in other departments of learning, especially the law.

The desire of the boys to advance grew with their stock of knowledge. Accordingly, when, in 1794, Hans was only seventeen years of age and his brother sixteen, they both matriculated at the University of Copenhagen. Their father was able to help them but little, so that they were obliged to live quietly and sparingly, a condition distinctly favorable to consecutive and efficient study. They became so successful in their pursuits that they soon began to attract attention. Having passed creditable examinations, they were recommended for pecuniary assistance from an educational fund established by the government for the purpose. Even then, as receipts were hardly equal to expenses, they sought to increase their little revenue by giving private lessons in their leisure hours. Here we have a striking example of what may be accomplished by men who work their way through College in the teeth of adverse circumstances; in these two brothers, we have proof of the truth that it is the student's mind, his willingness and determination to work, that count in education more than the golden opportunities that may fall to his lot.

In the year 1799, Oersted prepared a thesis on "The Architectonics of Natural Metaphysics," which won for him his Doctorate in Philosophy. Though the young Doctor did not hesitate to discuss metaphysical problems and even to disagree with Kant at a time when most Teutonic minds were deeply under the influence of the philosopher of Königsberg, his chief interests, however, centered in the experimental sciences, in physics and chemistry.

In spite of his devotedness to science, Oersted allowed himself, by way of distraction, an occasional excursion into the field of literature. A great literary and artistic movement was making itself felt in the northern part of Europe at the time. The æsthetic awakening of the Teutonic nations had come after three centuries of religious and political unrest, ill adapted to intellectual development. Lessing and Winkelmann, Goethe and Schiller, the two Schlegels and Klopstock as well as the young poets, Uhland and Koerner, were either already at work or were about to enter on their distinguished careers, and the neighboring Scandinavian nations were beginning to be seriously affected by the movement which was going on among their brethren. In the third year of his university course, Oersted entered the lists as a competitor for literary honors on the question, "What are the Limits of Prose and Poetry?" and had the satisfaction of winning the gold medal offered for the contest. In spite of this episode, indicative of devotedness to the muses, Oersted passed a brilliant pharmaceutical examination; and in the following year succeeded in capturing another prize, this time for a medical essay.

After such a period of preparation, it might be expected that a brilliant career would open up for Oersted; but, unfortunately, he could not afford to wait for slow academic rewards, as it was absolutely necessary for him to set about earning his livelihood. For this purpose, shortly after graduation, he accepted the position of manager of a drug-store. As the salary attached to the office was rather slender, he increased his resources by giving lectures in the evening on the familiar subjects of chemistry, natural philosophy and metaphysics.

About this time, the *wanderlust*, or passion for travel, took possession of our young philosopher; and under its influence, he resolved to see for himself what men of scientific avocations were doing in France and in Germany. His own pinched circumstances would not allow him to undertake such a journey; but he was fortunate enough to win a *stipendium cappelianum* which allowed him to travel at the expense of the government for a period of five years, though he used it only for three. If ever pecuniary aid was productive of enduring results, it was so in this case.

In 1801, at the age of twenty-four, Oersted set out from Copenhagen on his grand tour, determined to make it a scientific as well as sentimental journey. In Germany, which he first visited, he met Klaproth, the orientalist; Werner, the mineralogist; Olbers the astronomer; the philosophers Fichte, Schelling and the two Schlegels; and above all, the young and brilliant physicist Johann Wilhelm Ritter, who discussed with him the theory of the wonderful "pile" invented by Volta in the previous year, 1800.

In Paris, Oersted spent about fifteen months, during which time he was in habitual relations with many of the savants who were just then reflecting great lustre on French science. To mention but a few: there was Cuvier, the leading naturalist of his age; Abbé Haüy, crystallographer of world-wide reputation; Biot, the brilliant expounder of physics; Charles, the discoverer of the law which bears his name; Berthollet, the associate of Monge the mathematician, and Lavoisier, the chemist.

On his return to the Danish capital in 1804, Oersted delivered courses of lectures on electricity and magnetism, light and heat, before numerous and cultured audiences; and such was the success which he achieved that he was appointed, at the age of twenty-nine, to the chair of physics in the University of Copenhagen.

For nearly forty-five years he was destined to occupy this academical position, so that his connection with that seat of learning rounded out the full period of half a century.

While sedulously occupied with the duties of his chair and the pursuit of his favorite scientific subjects, Oersted was not unmindful of his civic and altruistic obligations. He frequently gave popular scientific lectures, which were open to women as well as to men. He helped in the organization of a bureau through which lectures would be given in various parts of the country, and thus became a pioneer in what we call to-day the university extension movement. When democratic ideas began to be discussed in Denmark after the French Revolution of 1830, Oersted was one of those who took part in the onward movement for the betterment of the people. In 1835, he coöperated in the foundation of the Society for the Freedom of the Press; and when Christian VIII. ascended the throne, he addressed the new monarch in a speech of liberal tendency, hailing him because of the interest which he took in the advancement of science and in the uplift of the masses.

An idea of the position accorded to Oersted by his colleagues in the world of science may be gathered from an address made by Sir John Herschel at the closing session of the Southampton meeting of the British Association in 1836, in which the distinguished astronomer said: "In science, there is but one direction which the needle will take when pointed towards the European continent, and that is towards my esteemed friend, Professor Oersted. To look at his cool manner, who would think that he wielded such an intense power, capable of altering the whole state of science, and almost the knowledge of the world? He has at this meeting developed some of those recondite and remarkable forces of nature which he was the first to discover, and which went almost to the extent of obliging us to alter our views on the most ordinary laws of energy and motion. He elaborated his ideas with slowness and certainty, bringing them forward only after a long lapse of time. How often did I wish to Heaven that we could trample down, and strike forever to earth, the hasty generalizations which mark the present age, and bring up another and safer system of investigation, such as that which marked the inquiries of our friend? It was in deep recesses, as it were, of a cell, that a faint idea first occurred to Oersted. He waited long and calmly for the dawn which at length broke upon him, altering the whole relations of science and life. The electric telegraph and other wonders of modern science were but mere effervescences from the surface of this deep, recondite discovery of his. If we were to characterize, by any figure, the usefulness of Oersted to science, we would regard him as a fertilizing shower descending from heaven, which brought forth a new crop, delightful to the eye and pleasing to the heart."

It may be noticed that in Oersted's day early specialization was fortunately unknown. His education was broad and his intellectual activities broader still. Quite as interesting as many of his scientific researches are some of his contributions to philosophy and some of his views on the significance of the material universe. Oersted, a man of the world with a wide range of interests and a philosopher who lived at high intellectual altitudes, was one of the all-round men in the history of thought who took active part in science, in literature, in politics and in social problems. He had the opportunity of meeting many of the renowned scientists and philosophers of the century, and had been very closely in touch with some of them. He was a regular attendant at scientific congresses, in which he distinguished himself by the leading part which he took in their deliberations. His opinions, therefore, on the great problems of life, religious, moral, social and political, challenge our respect even where they do not compel our approval. Our Danish philosopher deserves, then, to stand as the spokesman of his generation of savants on the great questions that concern man's relations to his fellow-men, to an all-wise Providence and to an enduring hereafter. His opinions on these matters are all the more interesting because they are in open contradiction with what is sometimes thought to be the views of scientists on such subjects.

One of the passages of his paper on "All Existence, a Dominion of Reason," contains some surprising anticipations of ideas that created a great stir in the intellectual world some fifty years ago. In 1846, that is, thirteen years before the publication of Darwin's "Origin of Species," Oersted discussed evolution and suggested explanations that are generally considered to have been forced from apologists when compelled to take up the work of reconciling Christian doctrines with scientific conclusions.

Writing in the middle 'forties, he said: "If we are now thoroughly convinced that everything in the material world is produced from similar particles of matter, by the same forces and in obedience to the same laws, we must allow that the planets have been formed according to the same laws as our own earth. They have been in process of development during immeasurable periods of time, and have undergone numerous transformations which have also influenced the vegetable and animal kingdoms of those remote periods. The lower forms of life advanced by gradual stages to higher and more complex states of organization, till at length (in a comparatively recent period) a self-conscious being was evolved, the crowning work of this long-continued process of development. Accordingly, we must allow a similar order of organic development to take place on the other planets of our solar family. There may be some which have not as yet attained the same degree of development that we have reached; but everywhere throughout the universe, creatures endowed with reason appear in due time, just as man appeared on our own globe. Their understanding is intimately connected with the organs of sense which they possess; therefore, the nature of their mental faculties cannot be essentially different from our own. That I may avoid even the appearance of materialism, I must direct attention to the conciliatory principle, that the natural environment from which man springs must be recognized as the work of the eternal, creative Spirit. In other words, our conception of the universe is incomplete, if not comprehended as a constant and continuous work of the eternally creating Spirit."

Thus far Oersted; let us here recall what Lord Kelvin, the representative scientist of his day, quoted with approval on a memorable occasion from the Danish scientist with regard to the basic truths of science, philosophy and religion. "It will not be foreign to our purpose if, called upon by the solemnities of this day, we endeavor to establish our conviction of the harmony that subsists between religion and science, by showing how the man of science must look upon his pursuits, if he understands them rightly, as an exercise of religion.

"If my purpose here was merely to show that science necessarily engenders piety, I should appeal to the great truth everywhere recognized, that the essence of all religion consists in love toward God. The conclusion would then be easy, that love of Him from whom all truth proceeds must create the desire to acknowledge truth in all her paths; but as we desire here to recognize science herself as a religious duty, it will be requisite for us to penetrate deeper into its nature. It is obvious, therefore, that the searching eye of man, whether he regards his own inward being or the creation surrounding him, is always led to the Eternal Source of all things. In all inquiry, the ultimate aim is to discover that which really exists and to contemplate it in its pure light apart from all that deceives the careless observer by only a seeming existence. The philosopher will then comprehend what, amidst ceaseless change, is the Constant and Uncreated, which is hidden behind unnumbered creations, the bond of union which keeps things together in spite of their manifold divisions and separations. He must soon acknowledge that the independent can only be the constant and the constant the independent, and that true unity is inseparable from either of these. And thus it is in the nature of thought that it finds no quiet resting place, no pause, except in the invariable, eternal, uncaused, all-causing, all-comprehensive Omniscience.

"But, if this one-sided view does not satisfy him, if he seeks to examine the world with the eye of experience, he perceives that all those things of whose reality the multitude feels most assured never have an enduring existence, but are always on the road between birth and death. If he now properly comprehends the whole array of nature, he perceives that it is not merely an idea or an abstract notion, as it is called; but that reason and the power to which everything is indebted for its essential nature are only the revelation of a self-sustained Being. How can he, when he sees this, be otherwise animated than by the deepest feeling of humility, of devotion and of love? If anyone has learned a different lesson from his observation of nature, it could only be because he lost his way amidst the dispersion and variety of creation and had not looked upwards to the eternal unity of truth."

As already said, Oersted lived to celebrate the fiftieth year of his connection with his university. This was in November, 1850, on which occasion his friends, pupils and the public generally united together in honoring him as a professor whose warm and animated lectures enraptured audiences; as a leader in the scientific advance of the times; and as a Christian to whom nature was but a manifestation of the Deity's combined wisdom and creative power.

The aged scientist, much touched by this popular demonstration as well as by the tokens of esteem given him by the King, spoke of this jubilee celebration as the happiest day of his life. The reader will recall another great man, great in the world of politics and great on the field of battle, who said that the happiest day of *his* life was that of his first communion.

A few months after celebrating his golden jubilee, Oersted passed away, after a short illness, on March 9th, 1851, deeply mourned by all.

Oersted was eminent as a scholar and equally eminent as a man; lenient in his judgment of others, he was strict with regard to himself; simple in his ways and frugal in living, he was benevolent to others, being always ready to give a helping hand wherever needed. To such a man may well be applied these beautiful words with which Priestley begins his "History of Electricity": "A life spent in the contemplation of the productions of divine power, wisdom and goodness, would be a life of devotion. The more we see of the wonderful structure of the world and of the laws of nature, the more clearly do we comprehend their admirable uses to make all percipient creation happy, a sentiment which cannot but fill the heart with unbounded love, gratitude and joy."

A statue to the memory of Oersted was unveiled in Copenhagen on September 25th, 1876, in presence of the King of Denmark, the King of Greece, the Danish Crown Prince and members of the Royal family, as well as numerous high officials, representatives of learned societies and a vast body of students and people assembled together to do honor to a man who was distinguished alike by his scientific attainments and philosophical acumen, and who, during his long life, never faltered in his devotedness to the welfare of his country as he never weakened in his defense of the great truths of religion.

CHAPTER VIII. ANDRÉ MARIE AMPÈRE.

Few men of the nineteenth century are so interesting as André Marie Ampère, who is, as we have seen, deservedly spoken of as the founder of the science of electro-dynamics. Extremely precocious as a boy, so that, like his immediate predecessor in discovery, Oersted the Dane, his rapid intellectual development drew down upon him ominous expressions from those who knew him, he more than fulfilled the highest promise of his early years. His was no one-sided genius. He was interested in everything, and his memory was as retentive as his intellect was comprehensive. He grew up, indeed, to be a young man of the widest possible interests. Literature never failed to have its attraction for him, though science was his favorite study and mathematics his hobby. The mathematical mind is commonly supposed to run in very precise grooves, yet Ampère was always a speculator, and his speculations were most suggestive for his contemporaries and subsequent generations. Indeed, his mathematics, far from being a hindrance to his penetrating outlook upon the hazier confines of science, rather seemed to help the penetrations it gave. While he was so great a scientist that Arago, so little likely to exaggerate his French contemporary's merit, has said of Ampère's discovery identifying magnetism and electricity, that "the vast field of physical science perhaps never presented so brilliant a discovery, conceived, verified, and completed with such rapidity," his friends knew this great scientist as one of the kindliest and most genial of men, noted for his simplicity, his persuasive sympathy and his tender regard for all those with whom he was brought into intimate relations.

The commonly accepted formula for a great scientist, that he is a man wrapt up in himself and his work, enmeshed so completely in the scientific speculations that occupy him that he has little or no time for great humanitarian interests, so that his human sympathies are likely to atrophy, is entirely contradicted by the life of Ampère. He was no narrow specialist, and, indeed, it may be said that not a single one of these great discoverers in electricity whom we are considering in this volume was of the type that is sometimes accepted as indicative of scientific genius and originality. After reading their lives, one is prone to have the feeling that men who lack that wider sympathy which, in the famous words of the old Latin poet, makes everything human of interest to them, are not of the mental calibre to make supreme discoveries, even though they may succeed in creating a large amount of interest in their scientific speculations in their own generation. It is the all-round man who does supreme original work of enduring quality.

André Marie Ampère was born at Lyons, January 22d, 1775. His father, Jean Jacques Ampère, was a small merchant who made a comfortable living for his family, but no more. His father and mother were both well informed for their class and time, and were well esteemed by their neighbors. His mother especially was known for an unalterable sweetness of character and charitable beneficence which sought out every possible occasion for its exercise. She was universally beloved by those who knew her, and the charm of Ampère's manner, which made for him a friend of every acquaintance, was undoubtedly a manifestation of the same family strain.

Shortly after the birth of their son, the parents gave up business and retired on a little property situated in the country not far from Lyons. It was in this little village, without any school-teacher and with only home instruction, that the genius of the future savant, who was to be one of the distinguished scientific men of the nineteenth century, began to show itself. For Ampère was not only a genius, but, what is so often thought to be an almost absolute preclusion of any serious achievement later in life, a precocious genius. The first marvelous faculty that began to develop in him was an uncontrollable tendency to arithmetical expression. Before he knew how to make figures, he had invented for himself a method of doing even rather complicated problems in arithmetic by the aid of a number of pebbles or peas. During an illness that overtook him as a child, his mother, anxious because of the possible evil effects upon his health of mental work, took his pebbles away from him. He supplied their place, however, during the leisure hours of his convalescence, when time hung heavy on his child hands, by bread crumbs. He craved food, but, according to the "starving" medical *régime* of the time, he was allowed only a single biscuit in three days. It required no little self-sacrifice on his part, then, to supply himself with counters from this scanty supply, and his persistence, in spite of hunger, evidently indicates that this mathematical tendency was stronger than his appetite for food. This is all the more surprising, since children are usually scarcely more than little animals in the matter of eating, and commonly satisfy their physical cravings without an after-thought of any kind.

Ampère learned to read when but very young, and then began to devour all the books which came to hand. Usually, the precocious taste for reading specializes on some particular subject; but everything was grist that came to the child Ampère's mental mill, and it was all ground up; and, strangest of all, much of it was assimilated. Travel, history, poetry, occupied him quite as much as romance; and, amazing as it may appear, even philosophy was not disdained while he was still under ten years of age. It seems amusing to read the declaration of the French biographer, that if this boy of ten had any special predilection in literature, it was for Homer, Lucan, Tasso, Fénelon, Corneille and Voltaire, yet it must be taken seriously.

When he was about fifteen, this omnivorous intellectual genius came across a French encyclopedia in twenty folio volumes. This seemed to him a veritable Golconda of endless riches of information. Each of the volumes had its turn. The second was begun as soon as the first was finished, and the reading of the third followed, and so on, until every one of the volumes had been completely read. References to other volumes might be looked up occasionally, but this did not distract him into taking other portions of the works out of alphabetical order. Surprising as it must seem, most of this heterogeneous mass of information, far from being forgotten at once, was deeply engraved on his wonderful memory. More than once in after-life, when many years had passed, it was a surprise to his friends to find how much information Ampère had amassed on some abstruse and unfamiliar subject, and how readily he was able to pour forth details of information that seemed quite out of his line. He would then confess that the encyclopedia article on the subject, read so many years before, was still fresh in his mind, or at least that its information was so stored away as to be readily available. We have heard much of Gladstone's memory in more recent years; but that seems to have been nothing compared to this wonderful faculty which recalled for Ampère, even as an old man, the unrelated details of every encyclopedia article that had passed under his eyes half a century before, when he was a boy of ten to fourteen.

The modest family library soon proved utterly insufficient to occupy the mind of this young, enthusiastic student; and his father, sympathetic to his ardent curiosity, took him to Lyons from time to time, where he might have the opportunity to consult volumes of various kinds that might catch his fancy. At this time, his old mathematical tendency reasserted itself. He wished to learn something about the higher mathematics. He found in a library in Lyons the works of Bernoulli and of Euler. When the delicate-looking boy, whom the librarian considered little more than a child, put in his request to the town library for these serious mathematical works, the old gentleman said to him: "The works of Bernoulli and Euler! What are you thinking of, my little friend? These works figure among the most difficult writings that ever came from the mind of man." "I hope to be able to understand them," replied the boy. "I suppose you know," said the librarian, "that they are written in Latin." This was a disagreeable surprise for young Ampère. As yet he had not studied Latin. He went home, resolved, however, to remove this hindrance to his study of the higher mathematics. At the end of the month, owing to his assiduity, the obstacle had entirely disappeared; and though he could read only mathematical Latin and had later to study the language from another standpoint, in order to understand the classics, he was now able to pursue the study of mathematics in Latin to his heart's content.

The even tenor of the boy's life, deeply engaged as he was in studies of every description, was destined to be very seriously disturbed. When he was but fourteen, in 1789, the Revolution came, with its glorious promise and then its awful consummation. Ampère's father was seriously alarmed at the revolutionary course things were taking in France, and had the fatal inspiration to leave his country home and betake himself to the city of Lyons. For a time, he occupied a position as magistrate. After the siege of Lyons, the revolutionary tribunal established there took up the project of making the Lyonnese patriotic, as they called it, by properly punishing the citizens for their failure to sympathize at first with the revolutionary government, and soon a series of horrible massacres began. New victims were claimed every day, and Ampère's father was one of those who had to suffer. The real reason for his condemnation was that he had accepted a position under the old government, though the pretext stated on the warrant for his arrest was that he was an aristocrat. This is the only evidence we have that the Ampère family was in any way connected with the nobility. The day on which he was sentenced to die, Jean Jacques Ampère wrote to his wife a letter of sublime simplicity, in which his Christian resignation of spirit, his lofty courage, yet thoroughly practical commonsense, are manifest. He warned his wife to say nothing about his fate to their daughter Josephine, though he hoped that his son would be better able to stand the blow, and perhaps prove a consolation to his mother.

The news proved almost too much for the young Ampère, and for a time his reason was despaired of. All his faculties seemed to be shocked for the moment into insensibility. Biographers tell us that he wandered around, building little piles of sand, gazing idly at the stars or vacantly into space, wearing scarcely any of the expression of a rational being. His friends could harbor only the worst possible expectations for him, and even his physical health suffered so much that it seemed he would not long survive. One day, by chance, Rousseau's "Letters on Botany" fell into his hands. They caught his attention, and he became interested in their charming narrative style, and as a result, his reason awoke once more. He began to study botany in the field, and soon acquired a taste for the reading of Linnæus. At the same time, classic poetry, especially such as contained descriptions of nature, once more appealed to him, and so he took up his classical studies. He varied the reading of the poets with dissections of flowers, and yet succeeded in following both sets of studies so attentively that, forty years afterward, he was still perfectly capable of taking up the technical description of the plants that he had then studied, and while acting as a university inspector, he composed 150 Latin verses during his horseback rides from one inspection district to another, without ever having to consult a gradus or a dictionary for the quantities, yet without making a single mistake. His memory for subjects once learned, was almost literally infallible.

Something of his love for nature can be appreciated from an incident of his early manhood, which is not without its amusing side. Ampère was very near-sighted, and had been able to read books all his life only by holding them very close to his eyes. This makes it all the more difficult to understand how he succeeded in reading so much. His near-sightedness was so marked that he had no idea of beauties of scenery beyond him, and was often rather put out at the enthusiastic description of scenes through which he passed *en diligence*, when his fellow-travelers spoke of the beauties of the scenes around them. Ampère, like most people who do not share, or at least appreciate, the enthusiasm of others for beautiful things around them, was in this mood, mainly because he was not able to see them in the way that others did, and, therefore, could not have the same pleasure in them. This lack in himself was unconscious, of course, as in all other cases, and, far from lessening, rather emphasized the tendency to be impatient with others, and rather made him more ready to think how foolish they were to go into ecstasies over something that to him was so insignificant.

One day, while Ampère was making the journey along the Saone into Lyons, it happened that there sat beside him on the stage-coach a young man who suffered from near-sightedness very nearly in the same degree as Ampère himself, but whose myopia had been corrected by means of properly fitting glasses. These glasses were just exactly what Ampère needed in order to correct his vision completely. The young fellows became interested in each other, and, during the course of their conversation, his companion suggested to Ampère, seeing how near-sighted he was, that he should try his glasses. He put them on, and at once nature presented herself to him under an entirely different aspect. The vision was so unexpected, that the description which he had so often heard from his fellow-travelers, but could not appreciate, now recurred to him, and he could not help exclaiming in raptures, "Oh! what a smiling country! What picturesque, graceful hills! How the rich, warm tones are harmoniously blended in the wonderful union of sky and mountain vista!" All of these now spoke emphatically to his delicate sensibility, and a new world was literally revealed to him. Ampère was so overcome by this unexpected sight, which gave him so much pleasure, that he burst into tears from depth of emotion, and could not satisfy himself with looking at all the beauties of nature that had been hidden from him for so long. Ever after, natural scenery was one of the greatest pleasures that he had in life, and the beauties of nature, near or distant, meant more to him than any other gratification of the senses.

In spite of the fact that Ampère had devoted considerable attention to acoustics as a young man, and had studied the ways in which the waves of air by which sounds are formed and propagated, he had absolutely no ear for music, and was as tone-deaf as he had been blind before his discovery with regard to the glasses. Musical notes constituted a mathematical problem for Ampère, but nothing more. This continued to be the case until about thirty years of age. Then, one day, he attended a musical soirée, at which the principal portions of the program were taken from Glück. It is easy to understand that this master of harmony possessed no charms for a tone-deaf young man. He became uneasy during the course of the musical program, and his uneasiness became manifest to others. After the selections of the German composer were finished, however, some simple but charming melodies were unexpectedly introduced, and Ampère suddenly found himself transported into a new world. If we are to believe his biographers, once more his emotion was expressed by an abundance of tears, which Ampère seems to have had at command and to have been quite as ready to give way to in public as any of Homer's heroes of the olden time. Blind until he was nearly twenty, he used to say of himself, he had been deaf until he was thirty. In spite of his failure to respond in youth, once it had been awakened to appreciation, his soul vibrated profoundly to all the beauties of color and sound, and, later in life, they gave rise in him to depths of emotion which calmer individuals of less delicate sensibilities could scarcely understand, much less sympathize with.

Between his two supreme experiences in vision and sound, there had come to Ampère another and even profounder emotion. He tells the story himself, in words that probably express his feelings better than any possible description of his biographer could do, and that show us how wonderfully sensitive his soul was to emotion of all kinds. He had just completed his twenty-first year when he fell head over heels in love. Though he wrote very little, as a rule, he has left us a rather detailed description in diaries, evidently kept for the purpose, of the state of his feelings at this time. These bear the title, "*Amorum*," the story of his love. On the first page these words occur: "One day as I was taking an evening walk, just after the setting of the sun, making my way along a little brook," then there is a hiatus, and he was evidently quite unable to express all that he felt. It seems that he was gathering botanical specimens, wearing an excellent set of spectacles ever since his adventure on the stage-coach had shown him the need of them, when he suddenly perceived at some distance two young and charming girls who were gathering flowers in the field. He looked at one of them, and he knew that his fate was sealed. Up to that time, as he says, the idea of marriage had never occurred to him. One might think that the idea would occur very gently at first, then grow little by little; but that was not Ampère's way. He wanted to marry her that very day. He did not know her name; he did not know her family; he had never even heard her voice, but he knew that she was the destined one.

Fortunately for the young lady and himself, she had very sensible parents. They demanded how he would be able to support a wife. Ampère was quite willing to do anything that they should suggest. His father had left enough to support the family, but not enough to enable him to support a wife in an independent home; and until he had some occupation, the parents of his bride-to-be refused to listen to his representations. For a time, he consented to be a salesman in a silk store in Lyons, in order to have some occupation which might eventually give him enough money to enable him to marry. Fortunately, however, he was diverted from a commercial vocation which might thus have absorbed a great scientist, and arrangements were made which permitted him to continue his intellectual life, yet have the woman of his choice. She was destined to make life happier far for him than is the usual lot of man, and he was ever ready to acknowledge how much she meant for his happiness.

With literature, poetry, love and settling down in life to occupy him, it is hard to think of Ampère as a young man doing great work in science, but he did; and his work deservedly attracted attention even from his very early years. It was in pure mathematics, perhaps, above all other branches, that Ampère attracted the attention of his generation. Ordinary questions he did not care for. Problems which the fruitless efforts of twenty centuries had pronounced insoluble attracted him at once. Even the squaring of the circle claimed his attention for a while, though he got well beyond it even before his boyhood passed away. There is a manuscript note from the Secretary of the Academy of Lyons, which shows that on July 8th, 1788, Ampère, then not quite thirteen years of age, addressed to that learned body a paper on the "Squaring of the Circle." Later, during the same year, he submitted an analogous memoir, entitled, "The Rectification of an Arc of a Circle, less than a Semi-circumference."

Arago says that he was tempted to suppress this story of Ampère's coquetting with so dangerous a problem, for Ampère rather flattered himself that he had almost solved it. It was only after Arago recalled how many geniuses in mathematics had occupied themselves with this same problem, that he saw his way clearly not to share the scruples of those who might think this incident a reflection on Ampère's mathematical genius. After all, Anaxagoras, Hippocrates, Archimedes and Apollonius, among the ancients, and among the moderns, Willebrod Snell, Huyghens, Gregory, Wallis, and finally Newton, the mathematician of the heavens, occupied themselves seriously with this very problem. Arago even notes that some men, by their speculations on the squaring of the circle, were led to distinguished discoveries, and mentions the name of Father Grégoire de Saint-Vincent, the distinguished Flemish mathematician of the Society of Jesus, to whom, as a direct result of his studies in attempted circle-squaring, we owe the discovery of the properties of hyperbolic space, limited by the curve and its asymptotes, as well as the expansion of log $(1 + x)$ in ascending powers of x. Montucla, the historian of mathematics, writing of Père Saint-Vincent, said that, "No one ever squared the circle with so much ability or with so much success." There was, however, a fallacy in his magnificent work which was pointed out by the celebrated Huyghens.

Shortly after the beginning of the nineteenth century, Ampère, as one of his French biographers rather characteristically declares, redeemed whatever of mathematical sinning there might have been, in indulging in fond dalliance with the squaring of the circle, by a series of mathematical papers, each of which was in itself a distinct advance on previous knowledge, and at the same time, definite evidence of his mathematical ability. The first paper, published in 1801, was a contribution to solid geometry, bearing the title, "On Oblique Polyhedrons." His next paper, written in 1803, though not published until 1808, was a treatise on the advantages to be derived in the theory of curves from due consideration of the osculating parabola. Another treatise, written about the same time, had for title, "Investigations on the Application of the General Formulæ of the Calculus of Variations to Problems in Mechanics." This concerned problems which had interested and, in most cases, proved too hard of solution even for such men as Galileo, Jacques Bernoulli, Leibnitz, Huyghens and Jean Bernoulli. Arago's expression with regard to this work is: "The treatise of Ampère contains, in fact, new and very remarkable properties of the *catenary* (la chainette) and its development." He adds: "There is no small merit in discovering hiatuses in subjects explored by such men as Leibnitz, Huyghens and the two Bernoullis. I must not forget to add that the analysis of our associate unites elegance with simplicity."

It is not surprising, after such marks of mathematical genius, that Ampère was appointed to the chair of mathematics at the École Polytechnique, where he came to be looked upon as one of the most distinguished of French mathematicians. In 1813, he became a candidate for the position left vacant by the death of the famous Lagrange; and at this time, presented to the Academy general considerations on the integration of partial differential equations of the first and the second order. After his election to the Academy, Ampère continued to present important papers at its various sessions. Among these, three are especially noteworthy: one was a demonstration of Père Mariotte's law (known to English students as Boyle's law); another bore the title, "Demonstration of a new Theory from which can be deduced all the Laws of Refraction, ordinary and extraordinary"; a third was a memoir on the "Determination of the curved surfaces of Luminous Waves in a medium whose Elasticity differs in each of the three dimensions."

In his eulogy of Ampère, which, together with his article in the "Dictionnaire Universelle de Biographie," we have followed rather closely, Arago calls particular attention to the fact that in Paris, Ampère moved in two intellectual circles quite widely separated in their interests and sympathies. Among the first group, were the members of the old "Institute" and professors and examiners of the École Polytechnique and professors of the Collège de France. In the other, were the men whose names have since become widely known as students of psychology, of whom Cabanis may be taken as the representative. Ampère had as great a passion for psychology, and was as ready to devote himself to fathoming and analyzing the mysteries of the mind, as he was to work out a problem in advanced mathematics, or throw light on difficult questions in the physical sciences. These two sets of interests are seldom united in the same man, though occasionally they are found. At the end of the nineteenth century, we had the spectacle of very distinguished men of science in physics, and even in biology—Sir William Crookes, Sir Oliver Lodge, Professor Charles Richet, Professor Lombroso and even Mr. Alfred Russell Wallace—interested in psychic and spiritualistic manifestations of many kinds as well as in natural science; and, inasmuch as they did so, they would have found Ampère a brother spirit. Ampère indeed dived rather deeply into what would be called, somewhat slightingly, perhaps, in our generation, metaphysical speculation. At one time, he contemplated the publication of a book which was to be called "An Introduction to Philosophy." He had made elaborate theories with regard to many metaphysical questions, and had written articles on "The Theory of Relations," "The History of Existence," "Subjective and Objective Knowledge" and "Absolute Morality." Arago calls attention to the fact that Napoleon's famous anathema against ideology, far from discouraging Ampère, rather seemed to stimulate him in his studies, and he declared that it would surely contribute to the propagation of this kind of speculation, rather than to its suppression. It was simply another case of Napoleon overreaching himself, though this was in the domain of ideas and not in the realm of politics, where his fate was to reach him some time later.

How deeply interested Ampère became in metaphysics will perhaps be best appreciated from the fact that, for progress in metaphysics, exercise in disputation is needed, and had been the custom in the old medieval universities. Ampère once made an arrangement to travel from Paris to Lyons and stay there for some time, provided a definite promise was made that at least four afternoons a week should be devoted to discussions on ideology. The journey to Lyons, a distance of two hundred and fifty miles, was no easy undertaking in those days. The Paris, Lyons and Mediterranean Express now whirls one down to the capital of the silk district in a night; but in Ampère's time, it took many days, and the journey was by no means without inconveniences, which were likely to be so troublesome that a prolonged rest was needed after it was over. Ampère seems quite to have exhausted the interest of his friends in Lyons, who found his metaphysical speculations too high for them, though they themselves were specializing in the subject and would be glad to tempt him into discussions of the exact sciences; but in lyrical strain he apostrophizes psychological studies: "How can I abandon the country, the flowers and running waters for the arid streets of the city! How give up streams and groves for deserts scorched by the rays of a mathematical sun, which, diffusing over all surrounding objects the most brilliant light, withers and dries them down to the very roots! How much more agreeable to wander under flitting shades, where truth seems to flee before us to incite us to pursue, than walk in straight paths where the eye embraces all at a glance!"

Had Ampère been less successful as a mathematician or an investigator of physical science, these expressions would seem little short of ridiculous. As it is, they provide food for thought. Ampère seemed to realize that, for the intellectual man, the only satisfaction was not in successful research so much as in application of mind to what promised results. As in everything else, it was the chase, and not the capture, that counted. Seldom has this idea been applied to intellectual things with so much force as it seems to have appealed to Ampère, and one is reminded of Malebranche's famous expression, "If I had truth in my hand, I would be tempted to let it go for the pleasure of recapturing it."

The principal source of Ampère's fame, however, for future generations, was to be in his researches in the science of electro-dynamics. The name of this science will ever be inseparably linked with that of Ampère, its founder. It was for that reason, of course, that the International Congress of Electricians decided to give his name to the unit of current strength, so that it has now become a household word, and will continue so for ages to come. In spite of the resemblances, much more than superficial, between magnetism and electricity, the identification of these two with each other seemed as yet very distant. It is curiously interesting, however, to note that Ampère himself, in a program of his course, printed in 1802, announced that the "professor will demonstrate that electrical and magnetic phenomena must be attributed to two different fluids which act independently of each other." Ampère's fame was to be founded on the direct contradiction of this proposition, which he proposed and triumphantly defended by a marvelous series of experimental illustrations eighteen years later. In the meantime, the discovery of another distinguished scientist, doing his work many hundreds of miles away, was to prove the stimulus to Ampère's constructive imagination, so as to enable him to fill out many obscure points of knowledge with regard to magnetism and electricity.

This suggestive discovery was that of Oersted, the sketch of whose life and work immediately precedes this. Oersted demonstrated that a current of electricity will affect a magnetic needle. This epoch-making discovery reached Paris by way of Switzerland. The experiment was repeated before the French Academy of Sciences by a member of the Academy of Geneva, on September 11th, 1820. The date has some importance in the history of science, for just seven days later, on the 18th of September, Ampère presented, at the session of the Academy of Sciences, a still more important fact, to which he had been led by the consideration of Oersted's discovery while testing it by way of control experiment. This brilliant discovery of Ampère, Arago summed up in these words: "Two parallel conducting wires attract each other when the current traverses them in the same direction. On the contrary, they repel each other when the current flows in opposite directions. The phenomenon described by Oersted was called, very appropriately, electromagnetic, whilst the phenomena described by Ampère, in which the magnet played no part, received at his suggestion the general name of electro-dynamics, which has since been applied to them."

At first it was said that these phenomena were nothing more than manifestations of the ordinary attractive and repelling power of the two forms of electricity which had been so carefully studied, especially in France, during the eighteenth century. Ampère at once disposed of any such idea as this, however, by pointing out that bodies similarly electrified repel each other, whilst those that are in opposite electrical states attract each other. In the case of conductors conveying currents, there is attraction when these are in the same direction, and repulsion when they flow in the opposite direction. This reasoning absolutely precluded all possibility of further doubt in the matter, and this particular form of objection to Ampère's discoveries was dropped at once.

Having satisfactorily disposed of other objections, Ampère was content neither to rest quietly in his discovery nor merely to develop various experimental phases of it which would be extremely interesting and popularly attractive, but which at the same time might mean very little for science. With his mathematical mind, Ampère resolved to work out a mathematical theory which would embrace not only all the phenomena of magnetism then known, but also the complete theory of the science of electro-dynamics. Needless to say, such a problem was extremely difficult. Arago has compared it to Newton's solution of the problem of gravitation by mathematics. Considering the comparatively small amount of data that Ampère had at his command, this problem might very well be compared to that which Leverrier took up with so much success, when he set about discovering by calculation only the planet Neptune, as yet unknown, which was disturbing the movements of Uranus.

It might be thought that these discoveries of Ampère would be welcomed with great enthusiasm. As a matter of fact, however, new discoveries that are really novel always have, as almost their surest index, the fact that contemporaries refuse to accept them. The more versed a man is in the science in which the discovery comes, the more likely is he to delay his acceptance of the novelty. This is not so surprising, since, as a rule, new discoveries are nearly always very simple expressions of great truths that seem obvious once they are accepted, yet have never been thought of. They mean, therefore, that men who consider themselves distinguished in a particular science have missed some easily discoverable phenomenon or its full significance, and so, to accept a new discovery in their department of learning men must confess their own lack of foresight.

It may be pointed out that the same thing happened with regard to Ohm, only it was much more serious. Years of Ohm's life were wasted because of the refusal of his contemporaries to accept his "law" at his valuation. Arago, in his life of Ampère, recalls that when Fresnel discovered the transverse character of waves of light, his observations created the same doubts and uncertainty in the same individuals who a few years later refused to accept Ampère's conclusions. Arago puts it, that as he was ambitious of a high place in the world of ideas, he should have expected to find his adversaries precisely those already occupying the highest places.

Ampère never looked on himself as a mere specialist in physical science, however, and it is extremely interesting to know that he dared to take sides in a discussion between Cuvier and Geoffroy-Saint-Hilaire, with regard to the unity of structure in organized beings. While the purely physical scientists mostly sat mute during the discussion, Ampère took an active share in it, and ventured to subject himself to what perhaps, above all things, a Frenchman dreads, the ridicule of his colleagues. Arago thought that he held his own very well in this discussion, which involved some of the ideas that were afterwards to be the subject of profound study and prolonged investigation later in the nineteenth century, because of the announcement of the theory of evolution.

After his discoveries in electricity Ampère came to be acknowledged as one of the greatest of living scientists, and was honored as such by most of the distinguished scientific societies of Europe. His work was not confined to electricity alone, however, and late in life he prepared what has been well called a remarkable work on the classification of the sciences. This showed that, far from being a mere electrical specialist or even a profound thinker in physics, he understood better probably than any man of his time the interrelations of the sciences to one another. He was a broad-minded, profound thinker in the highest sense of the words, and in many things seems to have had almost an intuition of the intimate processes of nature; a sharer in secrets as yet unrevealed, though he was at the same time an untiring experimenter, eminently successful, as is so evident in his electrical researches, in arranging experiments so as to compel answers to the questions which he put to nature.

In the midst of all this preoccupation of mind with science and all the scientific problems that were working in men's minds in his time, from the constitution of matter to the nature of life, above all engaged in experimental work, he was a deeply religious man in his opinions and practices. He had indeed the simple piety of a child. During the awful period of the French Revolution, he had some doubts with regard to religious truths; but once these were dispelled, he became one of the most faithful practical Catholics of his generation. He seldom passed a day without finding his way into a church, and his favorite form of prayer was the rosary.

Frederick Ozanam tells the story of how he himself, overtaken by misgivings with regard to faith, and roaming almost aimlessly through the streets of Paris trying to think out solutions for his doubts, and the problems that would so insistently present themselves respecting the intellectual foundations of Christianity, finally wandered one day into a church, and found Ampère there in an obscure corner, telling his beads. Ozanam himself was moved to do the same thing, for Ampère was then looked upon as one of the greatest living scientists of France. Under the magic touch of an example like this and the quiet influence of prayer, Ozanam's doubts vanished, never to return.

Saint-Beuve, whose testimony in a matter like this would surely be unsuspected of any tendency to make Ampère more Catholic than he was, in his introduction to Ampère's essay on the Philosophy of the Sciences (Paris, 1843), says:

"The religious struggles and doubts of his earlier life had ceased. What disturbed him now lay in less exalted regions. Years ago, his interior conflicts, his instinctive yearning for the Eternal, and a lively correspondence with his old friend, Father Barrett, combined with the general tendency of the time of the Restoration, had led him back to that faith and devotion which he expressed so strikingly in 1803.... During the years which followed, up to the time of his death, we were filled with wonder and admiration at the way in which, without effort, he united religion and science; faith and confidence in the intellectual possibilities of man with adoring submission to the revealed word of God."

Ozanam, to whose thoroughly practical Christianity while he was professor of Foreign Literatures at the University of Paris we owe the foundation of the Conferences of St. Vincent de Paul, which so long anticipated the "settlement work" of the modern time and have done so much for the poor in large cities ever since, was very close to Ampère, lived with him indeed for a while, said that, no matter where conversations with him began, they always led up to God. The great French scientist and philosopher used to take his broad forehead between his hands after he had been discussing some specially deep question of science or philosophy and say: "How great is God, Ozanam! How great is God and how little is our knowledge!" Of course this has been the expression of most profound thinkers at all times. St. Augustine's famous vision of the angel standing by the sea emptying it out with a teaspoon, which has been rendered so living for most of us by Botticelli's great picture, is but an earlier example of the same thing. One of Ampère's greatest contemporaries, Laplace, re-echoed the same sentiment, perhaps in less striking terms, when he declared that what we know is but little, while what we do not know is infinite.

For anyone who desires to study the beautiful Christian simplicity of a truly great soul, there is no better human document than the "Journal and Correspondence of Ampère," published some years after his death. He himself wrote out the love story of his life; and it is perhaps one of the most charming ot narratives, certainly the most delightful autobiographic story of this kind that has ever been told. It is human to the very core, and it shows a wonderfully sympathetic character in a great man, whose work was destined a few years later to revolutionize physics and to found the practical science of electro-dynamics.

When Ampère's death was impending, it was suggested that a chapter of the "Imitation of Christ" should be read to him; but he said, no! declaring that he preferred to be left alone for a while, as he knew the "Imitation" by heart and would repeat those chapters in which he found most consolation. With the profoundest sentiments of piety and confidence in Providence, he passed away June 10th, 1836, at Marseilles.

With all his solid piety, this man was not so distant from ordinary worldly affairs as not to take a lively interest in all that was happening around him and, above all, all that concerned the welfare of men. He was especially enthusiastic for the freedom of the South American Republics, eagerly following the course of Bolivar and Canaris, and rejoicing at the success of their efforts. South American patriots visiting Paris found a warm welcome at his hands, and also introductions that made life pleasant for them at the French capital. His house was always open to them, and no service that he performed for them seemed too much.

Ampère was beloved by his family and his friends; he was perhaps the best liked man among his circle of acquaintances in Paris because of the charming geniality of his character and his manifold interests. He was kind, above all, to rising young men in the intellectual world around him, and was looked up to by many of them as almost a second father. His charity towards the poor was proverbial, and this side of his personality and career deserves to be studied quite as much as what he was able to accomplish for science. The beauty of his character was rooted deeply in the religion that he professed, and in our day, when it has come to be the custom for so many to think that science and faith are inalterably opposed, the lesson of this life, so deeply imbued with both of these great human interests, deserves to be studied. Ozanam, who knew him best, has brought out this extremely interesting union of intellectual qualities, in a passage that serves very well to sum up the meaning of Ampère's life.

"In addition to his scientific achievements," says Ozanam, "this brilliant genius has other claims upon our admiration and affection. He was our brother in the faith. It was religion which guided the labors of his mind and illuminated his contemplations; he judged all things, science itself, by the exalted standard of religion.... This venerable head which was crowned by achievements and honors, bowed without reserve before the mysteries of faith, down even below the line which the Church has marked for us. He prayed before the same altars before which Descartes and Pascal had knelt; beside the poor widow and the small child who may have been less humble in mind than he was. Nobody observed the regulations of the Church more conscientiously, regulations which are so hard on nature and yet so sweet in the habit. Above all things, however, it is beautiful to see what sublime things Christianity wrought in his great soul; this admirable simplicity, the unassumingness of a mind that recognized everything except its own genius; this high rectitude in matters of science, now so rare, seeking nothing but the truth and never rewards and distinction; the pleasant and ungrudging amiability; and lastly, the kindness with which he met everyone, especially young people. I can say that those who know only the intelligence of the man, know only the less perfect part. If he thought much, he loved more."

CHAPTER IX. OHM, THE FOUNDER OF MATHEMATICAL ELECTRICITY.

Lord Kelvin, himself one of the greatest of the electrical scientists of the nineteenth century, in commenting some years ago on Ohm's law, said that it was such an extremely simple expression of a great truth in electricity, that its significance is probably not confined to that department of physical phenomena, but that it is a law of nature in some much broader way. Re-echoing this expression of his colleague, Professor George Chrystal, of Edinburgh, in his article on electricity in the Encyclopedia Britannica (IX. edition), says that Ohm's law "must now be allowed to rank with the law of gravitation and the elementary laws of statical electricity as a *law of nature* in the strictest sense." In a word, to these leaders and teachers in physical science of the generation after his, though within a comparatively short time after Ohm's death, there has come the complete realization of the absolutely fundamental character of the discovery made by George Simon Ohm, when he promulgated the principle that a current of electricity is to be measured by the electromotive force, divided by the resistance in the circuit. The very simplicity of this expression is its supreme title to represent a great discovery in natural science. It is the men who reach such absolutely simple formulæ for great fundamental truths that humanity has come, and rightly, to consider as representing its greatest men in science.

Like most of the distinguished discoverers in science who have displayed marked originality, Ohm came from what is usually called the lower classes, his ancestors having had to work for their living for as long as the history of the family can be traced. His father was a locksmith, and succeeded his father at the trade. The head of the family for many generations had been engaged at this handicraft. The first of them of whom there is any definite record was Ohm's great-grandfather, Wilhelm Ohm, who was a locksmith at Westerholt, not far from Münster, in Westphalia. Wilhelm Ohm's son, Johann Vincent, the grandfather of the great electrician, during his years as a journeyman locksmith had spent some time in France, and subsequently settled down in Kadolzburg, a small suburb of

Erlangen, in Bavaria. In 1764, he obtained the position of locksmith to the University of Erlangen, and became a citizen of that municipality. Both of his sons followed the trade of their father.

The elder of these, Johann Wolfgang, worked at his trade as a journeyman in a number of the small cities of Germany, and only after ten years of absence in what, because of the independent condition of the States now known as the German Empire, were then considered foreign parts, did he wander back to his native place. On his return he received the mastership in his craft, and shortly after, about 1786, married a young woman named Beck. George Simon Ohm, the electrical scientist, was the first child of this marriage, and was born March 16th, 1789. A second son, born three years later, also became distinguished in after-life for his mathematical ability. This younger brother, after having filled a number of teaching positions in various German educational institutions, was called as professor of mathematics to Berlin, where he died in 1862.

While their father, Johann Wolfgang Ohm, followed his trade of locksmith for a living, like many another handicraftsman, he had many mental interests which he cultivated in leisure hours, and doubtless dwelt on while his hands were occupied with the mere routine work of his trade. It is curiously interesting to find that he devoted himself, during the hours he could spare from his occupation, to two such diverse intellectual occupations as mathematics and Kant's philosophy; but they had no newspapers in those days, and a man, even of the artisan class, had some time for serious mental occupation. It might be thought, under these circumstances, that he would be but the most passing of amateurs in either of these subjects, and have a very superficial knowledge of them. This probably was true for his philosophy fad, for there are not many who have ever thought themselves more than amateurs in Kantism, and even Kant himself, I believe, thought that only one scholar ever really understood his system, and subsequently said he had some doubts even about that one; but in mathematics, the elder Ohm seems to have attained noteworthy success.

Hofrath Langsdorff, who was the professor of mathematics at Erlangen during the last decade of the eighteenth century, and who was called to Heidelberg in 1804, a fact that would seem quite enough to set beyond all question that his opinion in this matter may be taken as that of a competent judge, declared that the elder Ohm's mathematical knowledge was far above the ordinary, and that he knew much more than the elements even of the higher mathematics. Under these circumstances, it is not surprising that the father should have tried to encourage in both his boys a taste for mathematics, nor that he should have taken their mathematical instruction into his own hands and succeeded in making excellent mathematicians of them, even in their early years. He was so successful in this, indeed, that Langsdorff, after a five-hour examination of the brothers when they were respectively 12 and 15, did not hesitate to declare that the Erlangen locksmith's family was likely to be remembered as containing a pair of brothers who, for success in mathematics, might rival the famous Bernoulli brothers, so well known at that time.

This might be thought only a bit of neighborly praise, meant to warm a father's heart, yet it seems indeed to have been given quite seriously. Certainly the event justified the prophecy. It is not surprising that, with such a forecast to encourage him, the father should have been ready to make every sacrifice to enable both his sons to prepare for the university.

He continued his instruction of them, then, in mathematics, though he insisted at the same time that they should continue to keep up their occupation of locksmiths. In spite of his enthusiasm for mathematics, the old gentleman seems to have cherished no illusions with regard to the likelihood of pure mathematics ever serving them as a lucrative means of livelihood. It was a very satisfying intellectual interest, but a good trade was much more apt to prove their constant and substantial standby, unless, of course, the boys should actually prove to be the geniuses foretold. He seems to have realized to the full, Coleridge's idea that, like the literary man, the mathematician should have some other occupation, though he might not go to the extent of following Oliver Wendell Holmes' well-known addition to

Coleridge's formula, that he should, as far as possible, confine himself to the other occupation. The boys were given the opportunity to attend the gymnasium of Erlangen, and seem to have had excellent success in their general studies besides mathematics.

In 1805, when George, the subject of our sketch, was sixteen years of age, he was graduated from the gymnasium and was ready for the university. On May 3d, 1805, he took his matriculation examination before the faculty of Erlangen, electing the course of mathematics, physics and philosophy. Later in life he told his friends that it was his deep love for the mathematics of these studies, and his persuasion that in them the student was brought in contact with the most important factors for absolute intellectual cultivation, that tempted him to take them up. To this he did not hesitate to add that there seemed to him to be some call of a higher voice, as if he had a vocation to dedicate himself to the cultivation and extension of these important subjects.

He had been but some two years at the university, when for a time his studies had to be interrupted, partly for lack of means to pursue them, but partly because to his father, at least, the university course was not the source of such satisfaction as he had anticipated from his son's ability in mathematics. While Ohm took his studies seriously, he was not by any means a mere "grind," and, indeed, the reputation which he acquired at the university for many of the qualities which make for a student's popularity among his fellows, was not such as would be likely to appeal to a very serious-minded father. Ohm had acquired the fame of being one of the best dancers in the university; he was a brilliant billiard player and an unrivalled skater; all of which indicates that as a young man he had the physical development and acuteness of sense so necessary to enable him to gain prestige in all these sports.

His father, in spite of his desire for his son's university career, was quite willing, then, at the end of September, 1808, to have him take up a position as teacher of mathematics in the school kept by Pastor Zehnder, in the Canton Berne, in Switzerland. His very youthful appearance (he was only 18 years of age at the time, quite boyish looking and not even large for his years) caused the head of this institution no little surprise when he came with letters of introduction showing that he was to be the new teacher in mathematics. He could scarcely believe his eyes for a time. Within a few months, however, he was convinced of the ability and the capacity for work of his new addition to the faculty, who seems to have given, from the very beginning, excellent satisfaction in his rather important position.

Ohm remained there some three years and a half and then moved to Neunberg, where, independent of any educational institution, he set himself up as a private tutor in mathematics. His reason for so doing, as he himself tells, was that he wished to devote himself to the study of pure mathematics more than was possible in a regular teaching position. For this same reason also he refused a number of offers of positions as teacher of mathematics, which would ordinarily be considered quite flattering to a young man of only 21. Another reason for refusing these offers was that he wished to perfect himself in French, and he had an excellent opportunity afforded him for conversation in this language in the conditions in which he was placed in Neunberg. This last may seem an unusual reason, but it is characteristic of Ohm's determination always to add to his power of understanding and expression.

Most young men in Ohm's circumstances are so occupied with the thought of immediate success in life, that every possible abbreviation of their studies which will bring them nearer the opportunity to make their own living is likely to be heartily welcomed. Ohm, however, realized that his own intellectual development was more important, especially at this time, even than getting on in the world; and for this reason his life has an added interest, not only for students themselves, but especially for those who have the best interests of students at heart and wish to be able to cite examples of how a little delay in getting at one's actual life-work, or, still more, at a remunerative occupation, may serve the very useful purpose of preparing a man so much the better to bring out his best intellectual possibilities when he does settle down to his work.

At Easter, 1811, Ohm returned to Erlangen, after having spent nearly two years perfecting himself in mathematics. He then finished his studies at the university, which seems not to have had the rule of requiring attendance for a definite period before coming up for its degree, but permitted him to take the examinations for the doctorate of philosophy on the strength of the work he had done, and gave him his degree on the 25th of October of the same year. With the drawing tighter of the bands of red tape in educational institutions in more recent years, Ohm would have found it difficult to get his degree thus readily, though it was the university rather than the graduate who was eventually to be honored by it. After this, he became *privatdocent* in mathematics at the university, and taught for three semesters. He met with marked success and became very popular with the students. After a year and a half, however, he gave up his university position to accept the professorship of mathematics at the Realschule of Bamberg.

While Ohm was here, the spirit of young Germany awoke at the news of Napoleon's unfortunate Moscow campaign, in which his good fortune seemed to have definitely abandoned the great Emperor of the French. Most of the students of the universities of Germany were deeply aroused by it, and those who know Körner's and Uhland's songs will have some idea of the depth of patriotic feeling that was stirred in thousands of young German hearts, who thought that now the opportunity for the fatherland to throw off the hated foreign yoke forever, had come at last. Ohm debated with himself whether he should volunteer with the crowds of young men who were so bravely giving up everything, that the fatherland might be free. Two things deterred him. If he went as a soldier, the material assistance he was able to give his father, and which, as the old man was now advancing in years and had spent most of his little savings upon his sons, was needed, would have to be given up. The other motive that kept him at home was, according to his German biographer in the Allgemeine Deutsche Biographie, which we have been following for most of these details, because he felt that what he might be able to accomplish in other fields besides those of battle would eventually prove more beneficial for his fatherland, and indeed for the whole of humanity, than anything he could do as a soldier, even with the patriotic motive to help his country to throw off the yoke of the foreign usurper, which had proven so hard to bear. As we have already seen, it was a characteristic trait of Ohm all through life, that he cherished the idea, which acquired almost the force of a premonition, that he was destined for great things.

Ohm continued his work as a teacher, then, instead of volunteering for the army; but, as might be expected, found the monotonous work of drilling young students in mathematics extremely unsatisfactory after a time. At the end of a year and a half of service at Bamberg, he asked for a change in the conditions of his teaching position. Instead of this, he received a transfer to the Bamberg pro-gymnasium, where he was to teach Latin until a regular teacher was appointed. In spite of his representations that the teaching position offered him was utterly at variance with his talents and his inclinations, he was compelled to accept this occupation for a time, though after some delay there came the assurance that, just as soon as possible, he would be assigned to a position as teacher of mathematics.

In spite of his unfortunate circumstances, which would ordinarily be thought quite enough to keep him from serious work until he was settled in a position more suited to his tastes, he devoted himself to the writing of his first book during this time, and it was published by Enke, in Erlangen, in the spring of 1817. Its title was, "Outlines of the Study of Geometry as a Means of Intellectual Culture." It comprised nearly two hundred pages, and gives the best possible insight into the ability and intelligence of the author, then a young man of only twenty-eight. As a sort of appendix, he gives a short sketch of his father, evidently introduced, not quite so much for the purpose of filially confessing his obligations to the old locksmith mathematician, nor with the idea of repaying some of his immeasurable debt for all the opportunities which the sacrifices of paternal affection had brought into the life of his sons, as to emphasize the excellent educational influence which his father's mathematical training had had upon his boys, and thus prove his thesis as to the value of mathematical studies in education. Few filial tributes were ever more deserved or given more convincingly or with less suggestion of the conventional attitude of son to father.

Now that mathematics has come to occupy probably even a less prominent place in education than it did in Ohm's time, though the burden of his complaint with regard to educational methods was that geometry was not used as a daily developmental subject as much as it should be, it may be interesting to recall some of the reasons which he advanced for urging its greater employment as an instrument for mental training. He thought that rational geometry should occupy a place of honor among our means of education. Its quality as a mode of pure reasoning, though so closely related to the senses, made easy the transition from sensation to thought, which is such an important element in education; while its eminently simple character, though combined with definite demands upon the constructive faculties, made it appropriate in a high degree for the education of the young out of the field of merely imitative use of the intellect, into that of independent thinking and following out of ideas. "Geometry," says Ohm, "when properly taught, not with the fruitless drilling usually employed in teaching it, but in such ways as to secure deep personal attention, must take rank above all other branches of education, in enabling the student to break down the barrier which separates mere understanding from personal investigation. It forces a man whose thoughts were, up to this time, only the repetition of others' thoughts, to think for himself and to light for himself in his own mind the torches which enable him to see things clearly for himself, and not merely in the dimness of the half light that is thrown on them by the explanations of others."

Geometrical methods always had a special fascination for Ohm, and practically all of his books and writings bear the impress of that close dependence of all parts on one another, that absolutely logical connection so characteristic of geometric accuracy of thought. His was the sort of mind likely to be benefited by mathematical training. Such minds are, however, comparatively few, for most men are not rational in any sense of the word, that would make them dependent on logical reasoning. Perhaps it is as well that they are not, for many of those lacking in logic or mathematical accuracy of thought and absoluteness of conclusion, still continue to accomplish much in the world of thought and do much valuable planning for the complexities of human affairs, where strict logic will not always solve the intricate yet incomplete problems that present themselves in human relations, where, indeed, individual unknown factors often make any but an approximate solution impossible.

The opinions of the critics as to Ohm's "Outlines of Geometry" were, as might be easily anticipated, not all flattering, since only a few of the critics were able to place themselves on the ideal standpoint of mathematical subjectivity from which he had written his book. King Frederick William III., of Prussia, is said to have read it with much interest, however, and the royal pleasure doubtless drew attention to Ohm's work, and may have contributed to the fact that, shortly after its publication, in September, 1817, Ohm was invited by the Royal Consistory of Cologne to take the position of head professor of mathematics and physics in the gymnasium of that city. This post was not only honorable, it was also highly remunerative, at least from the standpoint of teachers' wages as they were at that time, and Ohm eagerly accepted the position.

Lamont, who was the director of the Royal Observatory at Munich, has written a memorial of Ohm which contains much valuable information. The body of it is an address delivered at a meeting of the Faculty of the University of Munich in honor of Thaddeus Siber and George Simon Ohm, but its value has been much enhanced by notes added before publication. Siber was a Benedictine who was professor in the philosophical department at Munich, and died the same year as Ohm. Lamont says that he received his information as to intimate details of Ohm's life from his brother, Prof. Martin Ohm, of Berlin. His sketch is, therefore, absolutely authoritative. Lamont says with regard to this period of teaching at Cologne: "Ohm's first position of importance, in any way worthy of his talents, was the professorship of mathematics at the large Jesuit gymnasium in Cologne, in 1817, where the special gift that he possessed, of making the study of mathematics not only comprehensible but attractive to boys, brought him success and recognition."

For nearly ten years Ohm had the opportunity to put into practice in this Jesuit gymnasium of the Rhineland, the principles which he had so much at heart, for he was apparently given the full freedom of his department of teaching. He succeeded so well that he received wide and hearty recognition for his work. The mathematical studies of the Cologne gymnasium stood higher than had ever been the case before, and this was all Ohm's work. In the years before his teaching in the Rhenish city, those who were distinguished in mathematics at the University of Bonn had not come, as a rule, from Cologne, but from other places; but now nearly all the mathematical prize-takers of Bonn came from among Ohm's students, and the best of the candidates for teaching positions in physics and mathematics had also, as a rule, had the advantages of his training.

Among the best of his scholars at this time was the afterwards well-known mathematician, Lejeune-Dirichlet, who taught in Berlin with Jacobi and Steiner and succeeded Gauss in Göttingen. Another of his most distinguished pupils was the astronomer Heis, who occupied a modest position at the Munster Academy, but whose merits were above the post which he occupied, and who was distinguished for the excellency of his original work and his ability as a mathematician. One very interesting fact with regard to Ohm's teaching, was that he was successful in catching and holding the interest not only of those of his students who were later to specialize in mathematics, but also of those who took up mathematics only as a subject for mental development, that was to be applied to other purposes later in life, and who found Ohm's teaching of the greatest possible service. Among these, the well-known German literary man, Jacob Venedey, of Cologne, has expressed his affection and gratitude for his old teacher in a very striking way in his sketch of the cathedral at Cologne, written in the banishment that came to so many vigorous German thinkers after the failure of the revolution of '48. In sending a copy of this to Ohm, Venedey says: "Honored Sir: — It will perhaps be a source of wonder to you that a student who apparently learned so little from you and your colleagues that he must now earn his bread by writing, should continue to cherish for you the liveliest gratitude. It is not the fault of mathematics that only the dimmest recollection of them remains with me. I shall never forget the personality of my professor, however, nor his ways and methods of teaching. I frequently recount your way with us boys, and I have the liveliest remembrance of your influence as a teacher. There are seldom weeks, there never is a month, when I fail to recall you. This is no mere compliment that I am paying to you, since I know you too well to think that flattery would mean anything to you, as it would be unworthy of you, and I for my part am not one of those who like to bandy compliments. I have often wished to meet you again, and a hundred times I thought that I saw you because some one at a distance had something that recalled you. I may say to you that you accomplished something for me in those days of teaching that I would not have been able to accomplish for myself. I can only think of you, then, with the highest feelings of reverence approaching what might well be called love. It will be a happy day, indeed, for me if I am ever in a position to make an hour of existence happier for you in any way."

While Ohm so zealously continued his instruction in both the upper classes of the gymnasium, he never lost from sight that higher aim of original research and investigation to which his genius disposed him.

His choice of a subject for original investigation wavered for a long time between mathematics and physics, but, as he himself declared, his experience having shown him that authority was prone to play a large role in mathematics, while the field was more open for personal research and observation in physics, he resolved to take up that department for his special studies, consoled by the idea that physics cannot be properly pursued without mathematics. Looking around to select a subject that would serve as a striking preface to his work in this department, though resolved at the same time to avoid one where he would be without rivalry, he found it all ready to his hand in what one of his contemporaries called the enigmatic phenomena of the galvanic current. This was to prove a fortunate selection, indeed, both for himself and the opportunity afforded his genius as well as for the science of electricity itself.

He then began a series of investigations, always experimental in character, and with the mathematical explanations of the phenomena observed carefully worked out. Accounts of these studies appeared from time to time in the year-book for Chemistry and Physics, issued by Schweigger. After some ten years, these were collected together, or at least the principal portions of them, and published in the second half of the year-book for the year 1826. The apparatus for his experiments was fortunately at command in the gymnasium at Cologne, but without his mechanical skill, obtained from his experience as a locksmith when a boy, it would have been impossible so to vary his experiments and modify his instruments as to bring out many of the phenomena that he succeeded in demonstrating. Nearly all of the great discoverers in science have been handy men possessed of mechanical skill, and this is as true for medicine, as I have shown in "Makers of Modern Medicine," though it might perhaps not be expected, as it is here in electricity, where it seems very natural.

Ohm felt, in 1826, that he had succeeded in exhausting nearly all that he could learn for himself, and as he wished to have opportunities for further study, and especially for further reading, he asked for an academic furlough that would carry him over the next year. The work that he had already accomplished was beginning to be appreciated, and after discussion of the papers that he had published up to that time, the requested furlough was promptly granted; and in a letter in which the school authorities praised his school work as well as his original investigations, they allowed him to take the sabbatic year for the furtherance of science on one-half the usual salary, though with the condition also that more would be allowed to him in case this seemed necessary and the conditions justified it.

This furlough was perhaps the most important event in Ohm's life. He employed it in bringing to a focus the ideas with regard to electricity which had been gradually worked out in his mind during the past ten years. In May, 1827, within six months after the beginning of his exclusive devotion to the subject, Ohm's article on the mathematics of the galvanic current appeared. It proved a scientific achievement of the first rank, that was to be epoch-making in the domain of electricity. It settled the conditions under which electrical tension exists in various bodies, and made it clear that there is a fundamental law of electrical conduction which could be expressed by an easy, simple formula.

Ohm's preface to his little book, that was to work such a revolution in electricity and was to remain for all time one of the classics in this department of science, is typical of the man in many ways. Its modesty could not very well be exceeded. Its simplicity constitutes in itself an appeal to the reader's interest. I know nothing in the literature of the history of science quite like it in these regards, unless it be the preface of Auenbrugger's little book on percussion, in which he laid the foundation of modern clinical diagnosis. The two men have many more qualities in common than the authorship of modest prefaces to their books. Both of them were geniuses whose names the aftertime will not willingly let die, and both of them accomplished their work apart from the stream of university life in their time, and met with a like fate in the neglect, for some time at least, by their distinguished colleagues of the important discoveries that they had made. Ohm's preface deserves to be quoted because of its classic quality:

"I herewith present to the public a theory of galvanic electricity as a special part of electrical science in general, and shall successively, as time, inclination and means permit, arrange more such portions together into a whole, if this first essay shall in some degree repay the sacrifice it has cost me. The circumstances in which I have hitherto been placed have not been suitable either to encourage me in the pursuit of novelties or to enable me to become acquainted with works relating to the same department of literature throughout its whole extent. I have, therefore, chosen for my first attempt a department of science in which I have the least to apprehend competition.

"May the well-disposed reader accept whatever I have accomplished with the same love for science as that with which it is sent forth! — The Author, Berlin, May 1st, 1827."

In his preface to the American edition of the "Galvanic Circuit Investigated Mathematically," Mr. Thomas D. Lockwood, vice-president of the American Institute of Electrical Engineers, said of this masterpiece of Ohm's: "A sufficient reason for republishing an English translation of the wonderful book of Professor G. S. Ohm is the difficulty with which the only previous translation (that of Taylor's Scientific Memoirs) is procurable.

"Besides this, however, the intrinsic value of the book is so great that it should be read by all electricians who care for more than superficial knowledge.

"It is most remarkable to note, at this time, how completely Ohm stated his famous law that the electromotive force divided by the resistance is equal to the strength of the current."

With regard to the book as a whole, Mr. Lockwood says, after suggesting certain anticipations of Ohm's ideas which had been made in the preceding century: "Ohm's work stands alone, and, reading it at the present time, one is filled with wonder at the prescience, respect for his patience and prophetic soul, and admiration of the immensity and variety of ground covered by his little book, which is indeed his best monument."

Like many another great discovery in physical science, Ohm's work failed to receive the immediate appreciation which it deserved. It cannot be said, however, that it failed to attract attention. It would be easier, indeed, to forgive the scientists of the day if this were true. Not long after its appearance, abstracts from it were made by Fechner in Leipzig, by Pfaff in Erlangen, and Poggendorff in Berlin, which showed that these scientists understood very clearly the significance and comprehended the wide application of Ohm's law as claimed by its author. From these men there was no question of hostile criticism. Professor Pohl, of the University of Berlin, however, in the Berlin

"Year-book of Scientific Criticism," did not hesitate to express his utter disagreement, and declared that Ohm's work was fallacious and should be rejected. Other writers of the time treated Ohm's article more or less indifferently, as a merely conventional contribution to science.

Professor Pohl's opinion was taken to represent the conclusions of the faculty of the University of Berlin, especially noted for mathematical ability. This was to prove a serious hindrance to Ohm in the university career which he had planned for himself. At Berlin they had the ear of the Minister of Education, and it was not long before Ohm felt that the criticisms of his work were making themselves felt in a direction unfavorable to him. Not long after the appearance of his book, there came a disagreement between Ohm and the educational authorities. Ohm felt that this was due to failure to recognize the significance of his work, and that under the circumstances he could not hope for the appreciation that would provide him with the opportunities he deserved. He insisted on sending in his resignation as a teacher. Nothing could change his determination in the matter, not even the pleas of his former scholars, and his resignation had to be accepted.

Ohm had hoped for a teaching position in a university. The Minister of Education declared that, while his work as a teacher had been accomplished with careful industry and diligence and conscientious attention to duty, the ministry regretted that, in spite of thorough appreciation of him and admiration for his excellent work as a scientist, they could not find for him a position outside of the gymnasium. How utterly trivial the conventional expressions sound, now that we know that they brought about for the time being the interruption of one of the most brilliant scientific careers in Europe. Of course, the geese cannot be expected to appreciate the swans, and it was not the minister's fault, but that of some of Ohm's own colleagues. The next six years of his life, the precious years between 38 and 44, Ohm had to give up the idea of teaching in a university, and devote himself to some private tutoring in Berlin, with a stipend of about three hundred dollars a year, miserable enough, yet sufficient, as would appear, for Ohm's simple mode of life. This he owed to the kindness of Gen. Radowitz, who employed him to teach mathematics in a military school in Berlin.

At the end of this time, when he was nearly 45 years of age, his unfortunate situation attracted the attention of King Ludwig I., of Bavaria, who offered him the chair of professor of physics at the Polytechnic School in Nuremberg, which had recently by royal rescript been raised to the status of a Royal Institute, with the same rank in educational circles as a lyceum for the study of humanities. Here Ohm's duties were shortly to be multiplied. He became the inspector of scientific instruction, after having occupied for some time the professorship of mathematics, and later became the rector of the Polytechnic School, a position which he held for some ten years, fulfilling its duties with the greatest conscientiousness and fidelity.

Ohm continued his work at Nuremberg for more than fifteen years. During this time, he succeeded in making his mark in every one of the departments of physics. He is usually considered as owing his reputation as an experimental and mathematical scientist to his researches in electricity. As a matter of fact, every branch of physics was illuminated by his work, and perhaps nothing shows the original genius of the man better than the fact that everything which he took up revealed new scientific aspects in his hands. The only wonder is that he should have remained so long in a subordinate position in the educational world at Nuremberg, and received his appointment as university professor of physics at Munich only in 1849.

In the midst of the administrative educational work that came to him at Nuremberg, Ohm did not neglect original investigation, but somehow succeeded in finding time for experiment and study. Having made a cardinal discovery in electricity, of the value of which surely no one was more aware than himself, Ohm might have been expected, as soon as his new post gave him the opportunity, to devote himself quite exclusively to this department of science. Instead, he turned for a time to the related subjects of sound, heat and light, devoting himself especially to their mathematics. He did this, as he said himself, to complete for his own satisfaction his knowledge of the scientific foundations of the imponderables, as heat, light and electricity were then called, but also because he wished, for the sake of his students, to get closely in touch with what had been accomplished by recent investigators in physics.

It is almost a universal rule in science, that no matter how distinguished an investigator may be, he makes but one cardinal discovery. Ohm, however, was destined, after having brilliantly illuminated electricity by the discovery of a great law, to throw nearly as bright a light on the domain of acoustics; and there is a law in this department of physics which is deservedly called by his name, though it is often associated with that of Helmholtz. Helmholtz himself was always most emphatic in his insistence on Ohm's priority in the matter, and constantly speaks of the law in question by Ohm's name.

Perhaps no better evidence of the breadth of Ohm's interest in science, his supreme faculty for experimentation, or the originality of his investigating genius, can be found than the fact that he thus discovered, by experimental and mathematical methods, the solution to important problems in two such distinct departments of physical science as electricity and acoustics. Before his time, the question of electrical resistance was absolutely insoluble. The problem in acoustics was not less obscure, as may be judged from the fact that, though some of the best physicists and mathematicians of Europe during the eighteenth century—and there were giants in those days, among others, Brook Taylor in England, D'Alembert in France, Johann Bernoulli and Euler in Germany, and finally, Daniel Bernoulli—had devoted themselves to its solution, it remained nevertheless unsolved. Here, as in electricity, the simplicity of the solution which Ohm found shows how direct were his methods of thinking and how thorough his modes of investigation. Perhaps the most striking feature of Ohm's work in acoustics, and, above all, his solution of an important problem in music, is the fact that he himself, unlike most of his German compatriots, had no ear for music and no liking for it.

In his address delivered at the public meeting of the Royal Bavarian Academy of Sciences at Munich, in March, 1889, the hundredth anniversary of the birth of Ohm, Eugene Lommel, in discussing the scientific work of Ohm, said: "Inasmuch as his law in acoustics furnished the clearest insight into the hitherto incomprehensible nature of musical tones, it dominates the acoustics of to-day no less completely than Ohm's law of the electric current dominates the science of electricity." This law concerns the resolution of tones into their constituents. The ideas laid down by Ohm were almost absolutely novel. They were so new that none of the workers in acoustics could think that Ohm had made a great discovery. His law states that the human ear perceives only pendulum-like vibration as a simple tone. Every other periodic motion it resolves into a collection of pendulum-like vibrations, which it then hears in the sound as a series of single tones, fundamentals and overtones. Ohm arrived at this law from mathematical considerations, making use of Fourier's series; for its experimental verification he was compelled to use the well-cultivated ear of a friend, inasmuch as he was himself, as we have said, quite devoid of musical appreciation.

Ohm's results were too distant from the accustomed ideas of investigators of sound at that time to be accepted by them. Seebeck, who was one of the most prominent scientists of the time in acoustics, did not hesitate to criticise severely, just as Pohl had made little of Ohm's law of the electric current. While, however, foreigners were to teach German scientists the value of the advance that their great colleague in electricity had made, the privilege of pointing out the significance of his work in sound was to be a compatriot's good fortune. It was nearly a score of years, however, before this vindication was to take place. Then Helmholtz, a decade after Ohm's death, furnished the experimental means which enabled even the unskilled ear to resolve a sound into its simple partial tones, and revolutionized the theory of music by his classic work, "The Science of the Perception of Sound," which is based entirely on Ohm's law of acoustics.

Ohm, in the appendix to his work, "The Galvanic Circuit treated mathematically," dared to suggest certain speculations with regard to the ultimate structure of matter. He said: "There are properties of space-filling matter which we are accustomed to look upon as belonging to it. There are other properties which heretofore we have been inclined to look upon as accidents or guests of matter, which abide with it from time to time. For these properties man has thought out causes, if not foreign, at least extrinsic, and they pass as immaterial independent phases of nature under the names light, heat, electricity, etc. It must be possible so to conceive the structure of physical bodies that, along with the properties of the first class, at the same time and necessarily those of the second shall be given."

It is all the more interesting to come upon Ohm's speculations on this subject of the ultimate constitution of matter, because within a few years of his time, Pasteur, then only a comparatively young man, had also been taken with the idea of getting at the constitution of matter by his observations upon dissymmetry, which he abandoned after a time, however, because he found other and more practical subjects to devote himself to, though he never gave up the thought that he might some time return to them and perhaps discover the underlying principles of matter from observations in this subject. It was not until the last five years of his life, when Ohm was already past sixty, that he was to enjoy the satisfaction of an ambition which he had cherished from his earliest years as a teacher, and which, in spite of untoward circumstances, had been a precious stimulus in his work. For some twenty years he had hoped some time to be able to devote himself to the investigation of the physical constitution of matter. Unfortunately, when the opportunity came, the manifold duties of his teaching position prevented the completion of his great work, and doubtless robbed his generation and ours of a precious heritage in the mathematics of the structure of matter, which would doubtless have been of the greatest possible value.

It is of course idle to speculate as to what he might have accomplished if left to his original investigation. The problem which he now took up was much more difficult than any of his preceding tasks. It would have seemed, however, quite as hopeless to those who lived before Ohm's laws, to look for a single complete law of the resistance of the electrical current in the circuit or of the overtones in music, as it is to us to think of a simple mathematical formula for atomic relations. What Ohm accomplished in these other cases by his wonderful power of eliminating all the unnecessary factors in the problem, would surely have helped him here. The main power of genius, after all, is its faculty of eliminating the superfluous, which always obscures the real question at issue to such a degree for ordinary minds, that they are utterly unable to see even the possibility of a simple solution of it. Art has been defined as the elimination of the superfluous; discovery in science might well be defined in the same terms. Under the circumstances, we cannot help regretting that Ohm was not allowed the time and the opportunity to work out the thoughts with which he was engaged. It would have been even more satisfactory if the precious years of his ripe middle age had not been wasted in trivial, conventional tasks, so that he might have been permitted to devote his academic leisure, sooner than was actually the case, to the problem which had been so constantly in mind since he made his great generalization in the laws of electricity.

Unfortunately, most of Ohm's time had now to be taken up with his teaching duties. Only for his self-sacrifice in the matter, his success as a teacher would doubtless have been less marked. Science itself must have suffered, however, from this pre-occupation of mind with a round of conventional duties, since Ohm could no longer devote his time to original research. In the meantime, his great discovery was coming to its own. During these ten years since the publication of his book, a number of distinguished physicists in every country—Poggendorff, and especially Fechner, in Germany, Jacobi and Lenz in Russia, Henry in America, Rosenkoeld in Sweden, and De Heer in Holland—took up the problems of the current strength of electricity as set forth in Ohm's law, and confirmed his conclusion by their investigations along similar lines. The French physicist and member of the Academy of Sciences, Pouillet, applied Ohm's ideas to thermo-electricity and pyro-electricity, employing his terms and bringing his work to the notice of foreigners generally, so that a translation of Ohm's work was made into English.

Ohm's work at once attracted the attention that it deserved in England. The Royal Society conferred on him the Copley Medal, which had been founded as a reward for important discoveries in the domain of natural knowledge. Before Ohm's time only one other German scientist, Carl Friedrich Gauss, of Göttingen, had ever been thus honored. The words employed by the Royal Society in conferring this distinction showed how thoroughly the representatives of English science appreciated Ohm's work. They said that he had set forth the laws of the electric current very clearly, and thus accomplished the solution of a problem which was as important in the realm of applied science as it had hitherto been in the schools. Recognition now became the rule, and Ohm had the satisfaction of having all his colleagues in the physical sciences acknowledge the significance of his work.

Ohm's recognition, then, came from foreigners first, and only afterwards from his fellow-countrymen. Immediate appreciation might have meant much for him, and even this tardy recognition gave him renewed courage and new strength to go on with his work. He gave effective expression at once to his gratitude and to the stimulus that had been afforded him by the dedication to the Royal Society of London of the great work, "Contributions to molecular Physics," which he planned.

The year after he received the Copley Medal, he was made a Foreign Associate of the Royal Society of England, and from this time on his discoveries began to find their way into text-books as fundamental doctrines in the science of electricity. German and foreign scientific bodies followed the English example so happily set for them, and began to give him their recognition as a physicist of the first rank. Ohm's further observations were, for a time, not accepted so readily as his first law. The reason for this was that Ohm was so far ahead of his times that there was not as yet in existence a suitable electroscope to test their truth. Finally, the invention of an exact electrometer by

Dellman, and its application by Professor Kohlrausch, of Marburg, made the experimental confirmation of all his work quite as significant as for his law.

It is a striking reflection on Ohm's career, though not very encouraging for the discoverer in science, to realize that some important discoveries, which thus proved eventually quite as epoch-making as his law, had lain for practically ten years neglected, and their magnificently endowed author had been allowed to eke out a rather difficult existence in teaching, not in the important department of science in which he was so great a master, but in certain conventional phases of mathematics which might very well have been taught by almost anyone who knew the elements of higher mathematics. Ohm's case is not a solitary phenomenon in the history of science, however, but rather follows the rule, that a genuine novelty is seldom welcomed by the leaders of science at any given moment; but, on the contrary, rather decried, and its discoverer always frigidly put in his proper place by those who resent his audacity in presuming to teach *them* something new in their *own* science.

Having thus illuminated electricity and acoustics, Ohm turned his attention to the department of optics. His power to simplify difficulties and get at the heart of obscure problems is illustrated by his contribution to this subject, made while he was professor of physics in the University of Munich. Optics had early engaged his attention, and in 1840 he published a paper in Poggendorff's Annalen, bearing the title, "A Description of some simple and easily managed Arrangements for making the Experiment of the Interference of Light." With his usual faculty for simplifying things, he showed that the interference prisms which were made so carefully by the French could be constructed from common plate-glass. He was indeed able to demonstrate that a simple strip from the edge of a piece of such glass could be used for this purpose.

He pursued this absorbing subject until 1852-53, and then set himself the difficult task of developing a general theory of these phenomena of interference which are so rich in form and color. The problem was indeed alluring, but some of the best minds in nineteenth century science in Europe had been engaged at it, without bringing much order out of the chaos, and it would have looked quite unpromising to anyone but Ohm, to whom, the greater the difficulty of a subject, the more the attraction it possessed. With his wonderful power of synthesis and his capacity to discover a clue to the way through a maze of difficulties, Ohm succeeded in finding a formula of great simplicity and beauty and which covered all the individual colors. It was only after he had reached his conclusions and was actually publishing his results, that the German scientist found that he had been anticipated by Professor Langberg, of Christiania, in Norway, with regard to the principal points of his investigation, though not as to all its details. Professor Langberg had published his article in the Norwegian Magazine for Natural Sciences in 1841, and an abstract of it had appeared the following year in the first complementary volume (Erganzungsband) of Poggendorff's Annalen.

Of this publication by Professor Langberg, Ohm had known absolutely nothing. He had even gone to some pains to find out, before undertaking his own investigation, whether anything had been published on the matter. At the sessions of the German Naturalists' Association, held in 1852, he had called the attention of many prominent physicists and mineralogists who were present at that meeting to the colored concentric ellipses which occur in connection with certain crystals used in the investigation of polarization. He asked whether these had ever been seen before, or whether anything had been written about them. All of those whom he consulted declared that they had not observed them, and that, so far as they knew, nothing had been published with regard to them. Accordingly, Ohm proceeded with his work, only to find, after its formal publication, that he had been almost entirely anticipated and that the merit of original discovery belonged to his Norwegian colleague.

When his attention was called to the publication, Ohm was perfectly ready to acknowledge the priority of Professor Langberg's claim and to give him all the credit that belonged to his discovery. At the beginning of the second part of his article, he said:

"I know not whether I should consider it lucky or unlucky that the extremely meritorious work of Langberg should have entirely escaped me and should have been lost to general recollection. Certain it is that, if I had had any knowledge of it before, my present investigations, which were occasioned by this elliptical system, would not have been made and I would have been spared a deal of work. In that case, however, a number of other and scarcely less important scientific principles would have remained hidden for the time being at least. Under the circumstances, the profound truth of the old proverb, 'Man proposes, but God disposes,' has been brought home to me again. What originally set me investigating this subject now proves to be without interest for science, since the problem has been solved before. On the other hand, a number of things of which I had no hint at all at the beginning of my researches, have come to take its place and compensate for it."

Perhaps nothing will show better than this, Ohm's disposition toward that Providence which overrules everything, and somehow, out of the mixture of good and evil in life, accomplishes things that make for the great purpose of creation. His eminently inquiring attitude towards science, which had on three occasions led him to tackle problems that had puzzled the greatest of experimental scientists, has been shown. He must have been, above all things, a man of a scientific turn of mind, in the sense that he was not ready to accept what had previously been accepted even by distinguished authorities in science, but was ready to look for new clews that would lead him to simpler explanations than any that had been offered before. In spite of this inquiring disposition, so eminently appropriate to the scientist, and constituting the basis of his success as an experimenter and scientific synthesist, he seems to have no doubts about the old explanation of the creation nor the all-wise directing power of a Divine Providence. This is all the more interesting, because already the materialistic view of things, which claims to know nothing except what can be learned from the matter around us, had begun to make its way in Europe, especially in scientific circles, but Ohm remained untouched by it.

Another example of this same state of mind in Ohm is to be found in the preface to his last great work, his contribution to molecular physics, in which he hoped to sum up all that he could discover and demonstrate mathematically with regard to the constitution of matter. He knew that he was taking up a work that would require many years and much laborious occupation of mind. He realized, too, that his duties as professor of physics and mathematics as well as the directorship of the museum and the consultancy to the department of telegraphs, left him comparatively little time for the work. He foresaw that he might not be able to finish it, yet hoped against hope that he would. In the preface to the first volume, he declared that he would devote himself to it at every possible opportunity, and that he hoped that *God would spare him to complete it*. This simplicity of confidence in the Almighty is indeed a striking characteristic of the man.

The work which Ohm began thus with such humble trust in God, was to contain his conclusions concerning the nature, size, form and mode of action of the atom, with the idea of being able to deduce, by the aid of analytical mechanics, all the phenomena of matter. Unfortunately, he was spared only to write the first, an introductory volume which bears the title, "Elements of the analytical geometry of space on a system of oblique co-ordinates." This did not touch, as he confesses, the ultimate problem he had in mind. The second volume was to have contained the dynamics of the structures of bodies, and a third and fourth were to be devoted to the physical investigation of the atom and its relation to other atoms and matter in general.

Ohm devoted himself, however, with too much ardor to his duties as teacher, to allow himself to give the time to his own work that would have enabled him to finish it. Among other things that he did for his students was to complete a text-book of physics. He confesses that he had always felt an aversion to working at a text-book, and yet was impelled to take up the task because he felt that in electricity, in sound and in optics, the only way in which his students would get his ideas, many of which were the result of his own work, was to have a text-book by himself, and he felt bound in duty to do this for them, as he had accepted the position of instructor. He succeeded in completing the book very rapidly by lithographing his lectures immediately after delivery and distributing copies to his classes.

It is almost needless to say that the work was, in its way, thoroughly original. It was accomplished with the ease with which he was always able to do things; but, unfortunately, the strain of the work told on him at his years much more than when, as a younger man, he was able to work without fatigue. He acknowledges, at the close of the preface, that the task has been too great, and that he should not have undertaken its accomplishment, and especially not in the hasty way in which it was done. This preface was dated Easter, 1854. Within a few months, Ohm's strength began to fail, and the end was not long in coming.

According to the translation of the address of Lommel, as it appeared in the Annual Report of the Smithsonian Institute for 1851, Ohm died as the result of repeated attacks of epilepsy, on July 6th, 1854. The date is correct; the mode of death, however, is surely reported under a misunderstanding. The physician who hears of epilepsy is prone at once to inquire as to its origin, and to wonder how long the patient had been suffering from it. There are no reports of previous attacks of epilepsy, and the sudden development of genuine epilepsy in fatal form at the age of 65 is quite unlikely.

His German biographer, Bauernfeind, who is quoted by Lommel as one of the authorities for the details of Ohm's life, and who was a pupil and intimate friend, gives quite a different account. Up to the very last day of his life, Ohm continued his lectures. His duties as professor appealed to his conscience as no others. On Thursday, July 6th, 1854, he delivered his last lecture. That night at ten o'clock he died. The cause of his death was given as a repeated apopleptic stroke. It is evidently because of the occurrence of more apopleptic seizures than one, that the assertion of epilepsy was introduced unto the account of his death.

For some days before his death, Ohm had been very weak, but had continued to fulfil every duty. To us in the modern time, it may seem surprising that there should be lectures in a university in July; but the second semester of the university year in Germany is not supposed to come to a close until the first of August, when the summer vacation begins, and lectures are continued until well on into July. The manner of Ohm's death, as told by his biographer friend, at once corrects the idea of epilepsy, and also shows that his passing came without any of the preliminary suffering that makes death a real misfortune. A half hour before his death, he had been entertaining some friends with lively recollections of the events of his early life in Cologne and Treves. He had been quite gay in the stories that he told, and almost boyishly happy in the recollections of those early days. For one for whom duty had meant so much in life, and who had always tried so faithfully to fulfil it, no happier call to higher things could possibly be imagined than that which came to Ohm.

On the following Sunday he was followed to the grave by numbers of friends, by all his colleagues and by most of the students of the Munich University. The university felt that it had suffered a great loss, and no signs of its grief were felt to be too much. Ohm was buried in the old Munich graveyard, where his bones still rest, beneath the simple memorial not unworthy of the modest scientist who did his work patiently and quietly, yet with never-failing persistency; who cared not for the applause of the multitude, and accomplished so much quite independently of any of the ordinary helps from others and from great educational institutions that are often supposed to be almost indispensably necessary for the accomplishment of original scientific work.

Ohm's personal appearance will be of interest to many of those to whom his discoveries have made him appeal as one of the great original thinkers in modern science. He was almost small in stature, even below middle height; and those who remember Virchow, may get something of an idea of his appearance when told that those who saw Ohm and knew Virchow, considered that there was a certain reminder of each other in the two men. According to his intimate friend and biographer, he had a very expressive face, with a high, somewhat doubled forehead. His eyes were deep and full of intelligence. His mouth, very sharply defined, betrayed, at the first glance, at once the earnest thinker and the pleasant man of friendly disposition. He was always restful and never seemed to be distracted. He talked but little, but his conversation was always interesting, and, except when he was in some particularly serious mood, was always likely to have a vein of light humor in it. He did not hesitate to introduce a sparkle of wit now and then into his lectures, and especially knew how gently to make fun of mistakes made by his pupils, yet in such a way as not to hurt their feelings, but to make them realize the necessity for more careful thought before giving answers, and for appreciating principles before speculating on them. He was particularly careful not to do anything that would offend his students in any way, and it is to this care that the success of his method of teaching has been especially attributed.

His habits of life were from the beginning of his career simple, and they continued to be so until the end. He was never married, and he himself attributed this to the unfavorable condition of his material resources at the beginning of his career as a teacher, and the fact that the improvement in these did not really come until he was well past fifty years of age. He once confessed to a friend that he missed those modest pleasures of family life which do so much to give courage and strength for the greater as well as the lesser sufferings of life. Most of his years of teaching he spent in boarding houses. Only after his appointment to the professorship at Munich was he able to have a dwelling for himself, which was presided over by a near relative.

Ohm is remembered as a teacher rather than as an educational administrator. His pupils recall him as one who was able to be eminently suggestive, while at the same time he succeeded in making it easy to acquire the details of information. The didactic lecture, as a method of teaching, did not appeal to him, and his success was due to the application of quite other methods. He realized how much personal influence meant, and the peculiarity of his system of teaching was an almost uninterrupted lively personal intercourse with his pupils. Demonstrations and exercises at the board always occupied the first half of his two-hour lesson, and only the other half was devoted to the setting forth of new matter. In this way, Ohm succeeded not only in influencing each student according to his personal endowments, but he also began the training of future teachers by giving them a living example of what their work should be.

The success of Ohm as a teacher was recognized on all sides. His attitude towards his scholars was very different from that which was assumed by many teachers. Instead of being a mere conveyer of scientific information, he was himself "a high priest of science," as one of his pupils declared, supplying precious inspiration, and not merely pointing out the limits of lessons and finding out whether they were known, but making work productively interesting, while neglecting none of the details. His pupils became distinguished engineers, and as this is the period in which the state railroads were being built, there was plenty of opportunity for them to apply the instruction they had received. Not only were the reports of the Royal Commission of Inspection repeated evidence of Ohm's success as a teacher, but the technical schools which were under the care of Ohm's disciples soon came to be recognized as far above the average, and as representing not only the successful teaching of technics on his part, but also the influence that his example as a teacher had in forming others to carry on the work.

How much Ohm was beloved by those who knew him best can be properly appreciated from the following passage from the panegyric delivered in Munich in 1855, not long after his death, by Professor Lamont, who had known him intimately: "Nature," he said, "conferred upon Ohm goodness of heart and unselfishness to an unusual degree. These precious qualities formed the groundwork of all his intercourse with his fellows. Despite the underlying strength of his character, which kept him faithfully at work during all his career, whenever there was question of merely personal advantage to himself, he preferred to yield to pressure from without, rather than rouse himself to resistance, and he thus avoided all bitterness in life. The unfortunate events which forced him, during the early part of his career, from an advantageous position back into private life, did not produce any misanthropic feelings in him, and when later a brilliant recognition gave him that rank in the world of science which by right belonged to him, his simplicity of conduct was not in any way modified, nor was the modesty of his disposition at all altered." In a word, Ohm was one of those rare geniuses whose magnanimity placed him above the vicissitudes of fortune. His power to do original work was not disturbed by the opposition which a really new discoverer invariably meets, but his unfailing equanimity was just as little exalted into conceit and pretentiousness by the praise which so justly came to him once the real significance of his scientific work dawned upon the world.

With the realization of all that Ohm's Work meant in the department of electricity, it is easy to understand how his name deserves a place in the science for all time. In order permanently to honor his memory, the International Congress of Electricians, which met at Paris in 1881, confirmed the action of the British Association of 1861, by giving the name *ohm* to the unit of electrical resistance. This is an ideal monument to the great worker. It is as simple and modest a reward as even he would have wished, expressing as it does, the gratitude of succeeding generations of scientists for all time.

CHAPTER X. FARADAY.

The maxim current among European scientists, that it is well to wait before accepting any scientific discovery to see what will be said about it on the other side of the Rhine, throws a rather curious sidelight on the supposed absoluteness of scientific knowledge. Gallic enthusiasm or German subtlety may evolve plausible theories that look like scientific discoveries, but the destructive criticism of the neighbor nation usually saves the scientific world from deception. Not infrequently, the English-speaking scientists held the balance between these rivals in the intellectual world, and their adhesion to either party or side of a question secured its dominance. When all three, Germans and French and English, are agreed as to the value of a scientific discovery, then it may be looked upon as having some of the absoluteness, or at least possesses for the moment the finality of scientific truth. If this triple agreement be taken as the criterion of the significance of a great scientist's work, then must Michael Faraday be considered as without doubt one of the greatest scientists of our time, and probably the greatest experimental scientist that the world has known.

Dubois Reymond, in Berlin, declared Faraday "the greatest experimentalist of all times, and the greatest physical discoverer that ever lived." Professor Martius said before the Academy of Sciences at Munich, "Deservedly has Faraday been called the greatest experimenter of his epoch, and that the greatest epoch of scientific experimentation down to our time." Dumas, the French chemist, in the panegyric delivered before the French Academy of Sciences, declared that Faraday was "the greatest scientific scholar that the Academy ever possessed." In order to give a picture of what he had accomplished in electricity, added Dumas, one would have to write a complete treatise on that subject. "There is nothing in this department of science that Faraday has not investigated completely or very materially modified. Much of this chapter of our modern science is his creation and belongs undeniably to him." Beside these testimonies from French and German scientific contemporaries must be placed Tyndall's appreciation, which sets forth his brother scientist's merits.

"Take him all in all," he said, "it must be admitted, I think, that Michael Faraday was the greatest experimental scientist that the world has ever seen."

Nor did these magnificent appreciations of Faraday cease when the enthusiasm for his memory, immediately after his death, had faded somewhat into sober realization of his merits. When Dumas summed up Faraday in the first Faraday lecture of the English Chemical Society, he said: "Faraday was a type of the most fortunate and the most accomplished of the learned men of our age. His hand, in the execution of his conceptions, kept pace with his mind in designing them; he never wanted boldness when he undertook an experiment, never lacked resources to insure success, and was full of discretion when interpreting results. His hardihood, which never halted once he had undertaken a task, and his wariness, which felt its way carefully in adopting a received conclusion, will ever serve as models for the experimentalist."

It is evident that the life of Faraday should be of supreme interest for a generation that is mainly interested in experimental science, and it so happens that his career contains many other sources of interest; for Faraday was a self-made man, who owed very little to anyone but himself and his own genius. Besides, he was a deep thinker with regard to all the problems of human life as well as those of science, and while he was a genial, kindly friend to those near him, the charming associate whom scientific intimates always welcomed, he had no illusions with regard to life being the end of all things, but looked confidently to the hereafter, and shaped his life here from that point of view.

Michael Faraday was born at Newington Butts, now called Stoke Newington, an outskirt of London, in Surrey, September 22d, 1791. His father was a journeyman blacksmith whose health was not very good, and as a consequence, the family suffered not a little from poverty. Both his parents were noted for their good habits, industrious lives and deep religious feelings. In spite of their poverty, as is much oftener the case than is sometimes thought, their children were brought up very carefully and had a precious training in high principles. Like most of his great colleagues in scientific discovery, Faraday had to begin to earn his livelihood early in life. Of educational opportunities he had practically none. He learned to read and write, and probably had a certain slight training in doing simple sums in arithmetic, but that was the extent of his formal teaching, and much of that he got at home. He had to help in the support of his family, and so it seemed fortunate that not far away from his home there was a bookstore and bindery, the owner of which became interested in the Faradays and took Michael as an errand boy when he was scarcely thirteen years of age.

It was here that the future scientist began his education for himself and, strange as it may seem, laid the deep foundation of his knowledge of science. For the first year he carried newspapers around to the customers, and did his work so faithfully that at the end of this time the book-binder offered to take him as an apprentice to the trade, without the usual premium which used to be rather strictly required for teaching boys their trades at that time. Faraday accepted this offer, but proved to be interested much more than in the outsides of the books he bound. Whatever of leisure there was he took advantage of to read a number of works on experimental science that happened to be in the shop. Luckily for him, some of these were classics. As an introduction to chemistry, he had Mrs. Marcet's "Conversations on Chemistry" and Robert Boyle's "Notes about the Producibleness of chimicall Principles." He was even more interested in electricity than in chemistry, however, and Lyons' "Experiments on Electricity" and the article on electricity in the Encyclopedia Britannica, whetted his interest and made the boy wish for more of such information. There probably could not be a better proof of the fact that, a man who really has intellectual interests will find the material with which to satisfy them, in spite of untoward circumstances, than this boyish experience of Faraday.

It is a curious anticipation of Faraday's after-career that he at once began to demonstrate by personal experiment some of the statements that he found in the books. He procured a stock of chemicals as far as his meagre salary would allow, and constructed a practical electrical machine, though he had nothing better than a large glass bottle to serve as a cylinder for it. When not yet fourteen, he noticed an advertisement of a set of lectures on natural philosophy. He was at once taken with the idea of going to them, but the price of admission, one shilling, seemed to place them entirely beyond him. His elder brother, who followed his father's trade of blacksmith, had more money than he, and, when properly cajoled, was persuaded to provide the necessary shillings, and so Faraday got to the lectures. Elder brothers do not often have to lend shillings to their juniors for admission to scientific lectures now any more than in Faraday's time, so that the incident seems worth noting.

In attendance at these lectures, Faraday not only learned much that was new to him in science, but met a number of earnest fellow-students and formed some life-long friendships. He took copious notes, and afterwards wrote them out in a fine, legible hand, making excellent drawings in perspective of the apparatus employed in the experiments. His notes were so extensive that Faraday bound them himself, in four volumes, with an index. These volumes are still preserved in the library of the Royal Institution as one of the precious treasures among its Faraday relics. The whole story of these early years of Faraday's life is a series of illustrations of how a young man without the necessary opportunities for his favorite studies can make them for himself. Everything seemed to be against his acquiring a thorough knowledge of science, yet he succeeded in creating for himself the equivalent of a good scientific course out of his meagre chances to hear lectures and read books on his favorite subject in the intervals of a busy life as book-seller and book-binder.

Things did not always continue to run along as pleasantly in life for young Faraday as while he was working for his book-binder friend as an apprentice. With the conclusion of his apprenticeship he became a journeyman book-binder, and his first employer proved to be a hard task-master. It did not matter how much work Faraday did or how well, it never quite satisfied this French émigré, until it is no wonder that Faraday looked for another occupation. For a time, he had the congenial occupation of acting as amanuensis for Sir Humphry Davy, who, while working on a new violent explosive, probably chloride of hydrogen, met with an accident which prevented him from using his eyes for some time. This occupation, pleasant and even alluring as it was, lasted only for a few days, however. It had the fortunate result of suggesting to Faraday to apply to Sir Humphry Davy in person for a position not long after, and it eventually brought him the position of assistant at the Royal Institution.

His anxiety to secure this post had been increased by the growing realization that a business life was not to his liking. It seemed to him a waste of time, or worse, for a man to give himself up to the making of money. Even thus young he had the ambition to add to the knowledge possessed by mankind, and the insatiable desire to increase the opportunities of others to learn whatever they were interested in. Accordingly, he set about finding the chance to devote himself entirely to science.

In writing years after to Dr. Paris, he says: "My desire to escape from trade, which I thought vicious and selfish, and to enter into the service of science, which I imagined made its pursuers amiable and liberal, induced me at last to take the bold and simple step of writing to Sir Humphry Davy, expressing my wishes, and a hope that, if an opportunity came in his way, he should favor my views; and at the same time I sent the notes I had taken of his lectures." Davy called, not long after, on one of his friends, who was at the time honorary inspector of the models and apparatus at the Royal Institution, and with the letter before him asked: "Here is a letter from a young man named Faraday; he has been attending my lectures and wants me to give him employment at the Royal Institution. What can I do?" "Do?" replied the inspector; "put him to wash bottles. If he is good for anything, he will do it directly; if he refuses, he is good for nothing." "No, no," replied Davy, "we must try him with something better than that."

Davy wrote a kind reply, and arranged for an interview with young Faraday. In this, however, he candidly advised him to stick to his business, telling him very plainly that "science was a harsh mistress, and, from a pecuniary point of view, but poorly rewarded those who devoted themselves to her service." He apparently put an end to all further consideration of the subject by promising Faraday the book-binding work of the Institution, and his own besides.

Faraday was not satisfied to go back to the book-shop, even with all this kindly patronage, but there was nothing else for it, and so for a time he continued at his duties and spent his spare moments reading science and his evenings at scientific lectures, or in remaking the experiments he had seen and others suggested by them, and above all in rewriting the notes that he had taken. There is no livelier picture in all the history of science, of how a man will, in spite of all obstacles, get the things he cares for, if he really cares for them, than that of Faraday thus teaching himself science in the face of what seems almost insurmountable discouragement. Fortunately, not long after he had been thus forcibly called to the attention of Sir Humphry Davy, the former assistant in the laboratory of the Royal Institution not only neglected his duties, but became a source of considerable annoyance. His misfortune proved Faraday's opportunity. He was offered the post. The salary was only twenty-five shillings a week, but he accepted it very willingly. One might think that at last his scientific career was opened for him, but his new post was no sinecure. The labors required from him, indeed, were so manifold that it is somewhat surprising that he found any time for his own improvement. His duties as set forth in writing were:

"To attend and assist the lecturers and professors preparing for and during lectures. Where any instruments or apparatus may be required, to attend to their careful removal from the model room and laboratory to the lecture room, and to clean and replace them after being used, reporting to the managers such accidents as shall require repair, a constant diary being kept by him for that purpose. That in one day in each week he be employed in keeping clean the models in the repository, and that all the instruments in the glass cases be cleaned and dusted at least once within a month."

The previous assistant had complained of the amount of work that was required of him. It is easy to see that his duties were rather exacting and time-taking. Faraday did not confine himself to them, though he did perform them with great assiduity. His interest in experimental chemistry was soon noted, and he was allowed to take his share in the experiments going on in the laboratory. Some of his first work was the extraction of sugar from beet-root; but he was soon to have abundant experience of the deterring side of chemistry. Not long after he began his work in the laboratory, he had to manufacture some bisulphide of carbon, one of the most nauseating of compounds. He found it disgusting enough as an experience, but the study of it brought its compensation.

It was much more than foul odors that Faraday had to encounter, for Davy was still occupying himself with the study of the explosives, in the investigation of which he had been injured the previous year. Faraday suffered from four or five explosions during the course of the first month or two of his employment. Indeed, the substance with which they were experimenting proved so unreliable in this regard that, after a second rather serious injury to Davy, further study of it was given up.

Once Faraday had secured his post at the Royal Institution, his life-work was before him, and he became deeply engaged in scientific speculations, investigations and experiments of all kinds. The young man who had found and made opportunities when they were so distant and difficult, now made use of all that were so ready at hand. He did not confine himself to his laboratory work, however, but seems always to have felt that the contact of minds engaged along the same lines was the best possible way to be stimulated to knowledge. He applied and was admitted as member of the Philosophical Society of London, an association of some two score of men occupied with many things during the day, but interested in science, so far as they could get the books and the opportunities for its study. They met every Wednesday evening and discussed various subjects in science or, as they called it then, in philosophy, and they seem to have occupied themselves with many questions in the social as well as the natural sciences. These men, most of whom were older than Faraday, soon came to look up to him because of the depth and increasing breadth of his knowledge, and we have some emphatic expressions of their admiration for him.

Faraday's earliest successful scientific investigation was accomplished in chemistry. This might have been expected, from the fact that he began his work with Sir Humphry Davy, whose principal scientific investigations had been concerned with chemistry. His own great scientific work was to be done in electricity. Even in the brief time that he devoted to chemistry, however, he succeeded in making some discoveries of deep significance. For instance, in his special study of chlorine, he demonstrated the existence of the two chlorides of carbon which had not hitherto been obtained. Above all, he impressed his personality upon methods in chemistry. He was the first to realize how much technics were to mean in the modern advancement of science, and he made methodic chemistry, in distinction from practical chemistry, the object of very special study. His work on *Chemical Manipulation* did more to train successful students of chemistry and to make good investigators in this department of science than any other single work in his generation. It has continued to be of interest down even to our own time, and is well worthy of consultation by all those who are interested in chemistry as a science, and especially in original research in that subject.

It was with regard to gases, however, that Faraday's most striking chemical work was done. He succeeded in liquefying several gases, and was the first to make clear that all matter could probably exist in each of the three different states—solid, liquid and gaseous—according as the proper conditions for each particular state were present. One might almost have expected that the serious dangers incurred in his early days in the Royal Institution, when his chief, Sir Humphry Davy, suffered so severely and he himself was more than once involved, might have deterred him from further investigation along similar lines; but Faraday's ardor for scientific investigation overcame any hesitancy there might have been. The effect of gases upon human beings proved as attractive to Faraday as it had been to Davy. His experiments upon chlorine threatened to prove seriously injurious to his throat, and he was warned of the danger that he was running in the effort to determine whether such gases were respirable and what their effects upon human beings were. The warning was disregarded, however, though he exercised somewhat more care in subsequent observations. His experiments in the respiration of gases finally led him to a discovery of cardinal importance in the very practical field of anæsthesia. Sir Humphry Davy, just at the beginning of the nineteenth century, had made a series of interesting experiments on nitrous oxide gas, the so-called "laughing gas," and had pointed out very definitely its anæsthetic properties. While suffering from toothache he had inhaled the gas, and had experienced prompt alleviation of the pain. He described in detail these curious effects, and suggested that there might be a place for nitrous oxide in surgery, at least for minor operations. The words he employed with regard to this subject show that the idea of anæsthesia, as we now understand it, had come to him very definitely. Not quite a score of years later, Faraday, recalling the experiments of Davy with nitrous oxide, studied sulphuric ether, and showed that the inhalation of the vapor of this substance produced anæsthetic effects very similar to those of nitrous oxide gas, but with the possibility of prolonging them much more easily and apparently with less danger than would be the case with the latter. In every history of anæsthesia, these two sets of experiments at the Royal Institution must be set down as foundation-stones, and Faraday's name particularly must be hailed as one of the initiators of a supremely beneficent advance in modern surgery.

Faraday had given up business to devote himself to science, and he was not to be seduced from the purpose of making his life unselfish and doing things, not for money, but for the good of science and his own satisfaction. As a practical chemist, he soon had many opportunities to increase his salary by making analyses for industrial purposes. During one year, the amount of work thus offered him was paid for so well that it formed an addition of some £500 sterling to his salary. It took away precious time, however, that he might otherwise devote to original work. As soon as Faraday realized this possibility of interference with his scientific investigations, he cut it off, quite content to live on the modest salary of his position at the Royal Institution. His action in the matter would remind one very much of Pasteur, in the latter half of the century, when asked by the Empress Eugénie, to whom he had been just exhibiting his discoveries in fermentation, whether he would not apply these to actual manufacture and so make a fortune for himself in brewing. Pasteur replied that he thought it unworthy of a French scientist to devote his time to money-making, with all the world of science open before him.

With a conscientious patriotism, however, that was typical of the man and his ways, there was one exception to this rule of not taking outside work that Faraday made. In a letter to Lord Auckland, long afterward, he says: "I have given up for the last ten years or more, all professional occupation and voluntarily resigned a large income, that I might pursue in some degree my own objects of research. But in doing this I have always, as a good subject, held myself ready to assist the government if still in my power, but not for pay; for, except in one instance (and then only for the sake of the person joined with me), I refused to take it. I have had the honor and pleasure of application, and that very recently, from the Admiralty, the Ordnance, the Home Office, the Woods and Forests and other departments, all of which I have replied to and will reply to as long as strength is left me."

As we have said, Faraday's principal work was accomplished in the domain of electricity. His supreme discovery, and, indeed, the most important practical discovery in the whole realm of electricity, was that of the induction effect of a current of electricity on a neighboring circuit. This was accomplished by experimental work of the highest order. Toward the end of 1824, when he was about thirty-three, he came to the definite conclusion that an electric current might be obtained by the motion of a magnet. His mind had been prepared for such a conclusion by Oersted's significant discovery in July, 1820, that an electric current acts somewhat like a magnet when the wire through which it flows is free to move. This discovery, definitely connecting electricity and magnetism, had been elaborated to an important degree by Ampère, and its sphere of application broadened by Wollaston. The curious though not unusual result in such cases, that it is not those who are in immediate touch with a great discoverer who develop or even apply his work, was illustrated by the fact that Ampère, the Frenchman, took up Oersted's discovery first, while Wollaston, working in England, had been the next one to follow successfully in the path thus opened up. It takes genius to go even a slight step farther into the unknown; the trained talent of disciples does not suffice. It was now Faraday, though not under Wollaston's influence, who was to continue successfully these labors.

In spite of his persuasion that a magnet would produce by induction an electric current, and the further step that a current in one wire could induce a current in another, experiments during seven years had brought him very little nearer the actual demonstration of this important principle. Those who think that great discoveries are made by accident and almost fall into the laps of their makers, as the apple upon Newton, should recall these seven years of unsuccessful labor on the part of Faraday. Finally, in 1831, he obtained the first definite evidence that an electric current can induce another in a different circuit. The discovery meant so much for him, that he hesitated to believe in his own success. Nearly a month after this first demonstration for himself, he wrote to his friend Phillips: "I am busy just now again on electro-magnetism, and think I have got hold of a good thing, but can't say. It may be a weed instead of a fish that, after all my labor, I may at last pull up."

He had long suspected, as we have said, that induction should occur, and he had tried currents of different strength, but without result. One day he noticed that, though he could not produce a permanent induced current, whenever the primary current started or stopped, there was a movement of the galvanometer connected with the secondary circuit, though the galvanometer remained at zero so long as the primary current flowed steadily. From this he proceeded to the demonstration that a bar magnet suddenly thrust into a helix of copper wire produced the same effect on the galvanometer, and evidently induced a transient current. When the magnet was withdrawn, the galvanometer needle swung in the opposite direction, showing another current, so that electrical currents were evidently induced by the relative motions of a magnet and a conductor. He continued his experiments in many different forms, and in the short space of a little more than a week, once the first definite hint was obtained, succeeded in so completely finding out the phenomena of electro-magnetic induction that scarcely more than practical applications in this subject were left for his successors.

Faraday's explanation of the induction of currents in the secondary circuit was probably quite as important a contribution to science as the series of experiments by which he demonstrated the occurrence of induced currents. His mind was not of the order that would accept action at a distance; that is, without some conducting medium through which the action took place. The old aphorism of the scholastics, "*actio in distans repugnat*"—action at a distance, that is, without a medium intervening, is absurd—would have appealed to him as a basic truth. The explanation that he outlined for induced currents was based on the lines of magnetic force, which he had so often delineated by means of iron filings. It was a favorite occupation of his, at moments of comparative leisure, to make varied pictures in iron filings of magnetic fields as they were exhibited under the influence of different combinations of magnets. He strewed iron filings over "gum paper," and then when the filings had arranged themselves in certain definite lines, he threw a jet of steam on the paper, which melted the gum and fixed the filings in position. He explained electrical action as the transmission of force along such lines as these, and he thought the whole electric field was filled with them.

Probably the best summary of Faraday's work on induction and its significance has been given us by Clerk Maxwell, in his article on Faraday, in the ninth edition of the Encyclopedia Britannica. There is no doubt but that Maxwell, above all men of the nineteenth century, was in a position to judge of the meaning of Faraday's work. He was not the sort of a man to say things in a panegyric mood, and his article on Faraday is indeed a model of well-considered judgment and critical illumination. Summing up the significance not only of Faraday's great discovery of induction, but also his theory in explanation of that discovery, he does not hesitate to say that his (Faraday's) opinion is the nearest approach to truth that has been advanced in this much-discussed subject.

"After nearly half a century of labor of this kind, we may say that, though the practical applications of Faraday's great discovery have increased and are increasing in number and value every year, no exception to the statement of these laws as given by Faraday has been discovered; no new law has been added to them; and Faraday's original statement remains to this day the only one which asserts no more than can be verified by experiment, and the only one by which the theory of the phenomena can be expressed in a manner which is actually and numerically accurate, and at the same time within the range of elementary methods of exposition."

With what eminent care and absolute truth Faraday's conclusions were reached may be judged from some further expressions of Clerk Maxwell's in the article just quoted, with regard to the attitude of certain mathematicians toward Faraday's work. In this matter, Clerk Maxwell, in talking on a theme that he had made especially his own, and in which his opinion must carry the greatest possible weight, said:

"Up to the present time, the mathematicians who have rejected Faraday's method of stating his law as unworthy of the precision of their science, have never succeeded in devising any essentially different formula which shall fully express the phenomena, without introducing the hypotheses about the mutual action of things which have no physical existence, such as elements of currents, which flow out of nothing, then along the wire, and finally sink into nothing again."

Faraday's results were described in papers afterwards incorporated in his first series of "Experimental Researches," which were read before the Royal Society, November 24th, 1841. These papers probably contain the best possible proof of Faraday's genius as an experimentalist and a leader in scientific observation. Within a few months after his first successful experiment, he had succeeded in bringing to perfection the whole doctrine of induction by currents and magnets, had laid down the fundamental ideas which were to constitute the formal basis of electro-magnetism for all time. Perhaps no better idea of the importance of the discovery thus made by Faraday can be given than will be found in Clerk Maxwell's compendious paragraph on this subject, in his sketch of Faraday, in the Encyclopedia Britannica. It may be said that no one in all the nineteenth century was more capable of appreciating properly the value of Faraday's work than this great electrical mathematician, who laid the firm foundation of mathematical electricity during the latter part of the nineteenth century. Clerk Maxwell says:

"This was of course a great triumph, and nobody appreciated this fact better than Faraday himself, who had been working at its problems for many years. One of the first problems that he had set himself in his note-book as a young man, was 'to convert magnetism into electricity,' and this he had now done. Within a month of the time that his first successful experiment was formed, he succeeded in obtaining induction currents by means of the earth's magnetism. Within a year he took the further immense step of obtaining a spark from the induced current. This would ordinarily have seemed quite impossible, since sparks occur only if the electromotive force is very high, and it was very low in his induced currents. He found, however, that if the circuit of wire in which a current was flowing is broken while the current is passing, a little bridge of metallic vapor is formed, across which the spark leaps. The difficulty with the experiment was to break the circuit during the extremely short period while the current is flowing. Faraday succeeded in doing this, and as a result obtained the first germ of the electric light. When he demonstrated this experiment by a very ingenious apparatus at the meeting of the British Association at Oxford, all were deeply interested, yet probably no one, even the most sanguine of the scientists present, thought for a moment that they saw the beginning of a far-reaching revolution of all the lighting of the world."

Perhaps the most interesting of Faraday's discoveries, from the scientific standpoint, because they throw so much light on the problems of all the related phenomena of magnetism, heat, light, even electricity, were those in which a ray of polarized light was used as a means of investigating the condition of transparent bodies when acted on by electric and magnetic forces. Faraday himself, when he was just thirty years of age, made a note in his commonplace laboratory book, in which all his observations were carefully detailed, that serves to show how much this subject had begun to interest him thus early in his career. He mentions that he had polarized a ray of lamp-light by reflection, and had made various experiments to ascertain whether any depolarizing action was exerted on it by water placed between the poles of a voltaic battery in a glass cistern, or by various fluids which were decomposed by the voltaic action during the course of the experiment. Besides water, the fluids used were weak solutions of sulphate of soda and strong sulphuric acid. None of them had any effect on the polarized light, either during the passage of the voltaic current or when this was shut off. No particular arrangement of particles in reference to polarized light could be found from these observations.

Such a note, with utter failure for conclusion, is common enough in Faraday's note-book. He was never discouraged, however, by failure at the beginning. Once a subject has been taken up seriously, it is almost inevitable that further observations with regard to it will be found during the course of the year. Because he had asked one question of nature and had not obtained a satisfactory answer, was never a reason why he should not ask further questions along the same line; and, above all, why he should not ask the same question in another way. After having tried a continuous current, Faraday next experimented on the effect of making and breaking the circuit. He did not expect very much from this, but he hoped that under circumstances when no decomposition would ensue as the effect of the current, he might find some indication of the polarization. It was nearly twenty-five years before Faraday succeeded in solving the problem that he had thus set himself as a young man, and nearly twenty years more were to pass before he made the relation between magnetism and light the subject of his very last experimental work. Nothing discouraged him. When he had resolved to investigate something, he continued to make his experiments over and over again in different ways, until finally he got an answer to his question and a solution to the problem.

Indeed, his perseverance in anything that he undertook was a striking characteristic of the man and one of the most important elements in his success in life. His tenacity of purpose showed itself equally in little as in great things. Arranging some apparatus one day with a philosophical instrument-maker, he let fall on the floor a small piece of glass. He made several ineffectual attempts to pick it up. "Never mind," said his companion, "it is not worth the trouble." "Well, but, Murray, I don't like to be beaten by something that I have once tried to do."

Faraday was sure that there was some very definite relation between electricity and light. His experiments, however, did not enable him to demonstrate this until nearly fifteen years after his successful experiment on induction. In September, 1845, he placed a piece of heavy glass made of silico-borate of lead in the field of a magnet, and found that, when a beam of polarized light was transmitted through the glass in the direction of the lines of force, there was a rotation of the plane of polarization. Later experiments showed him that all transparent solids and liquids were capable of producing this rotation in greater or less degree. When no magnet was used and the transparent substance was placed within a coil of wire through which an electric current was flowing, similar effects were produced. This was the demonstration of a definite relation between light and electricity. Later, Faraday found that magnets had a directive action upon the glass. He then made experiments upon gases, and found that they too exhibited magnetic phenomena, and that, indeed, the diurnal variations of the compass-needle were due to the sun's heat diminishing the magnetic permeability of the oxygen of the air. Further experiments with gases showed him that nitrogen was absolutely neutral in its reaction.

It might have been expected, from Faraday's early interest in chemistry, that when he turned to electricity and made discoveries in that field of research, he would naturally take up the problem of tracing the laws and demonstrating the relationships of the points of contact of the two great sciences. After his completion, then, of the subject of induction, Faraday devoted himself to the experimental proof of the identity of frictional and voltaic electricity, and to showing that chemistry and physics have a common ground. His inductive electrical machine could deflect a magnet and decompose iodide of potash. With his tendency to measure things, he determined that the amount of electricity required to decompose a grain of water was equal to 800,000 charges of his large battery of Leyden jars. On the other hand, the current from a frictional machine deflected the needle of his galvanometer in the same way as the induced current of electricity, so that all the elements of the proof of the identity of the two forms of phenomena were now in his hands.

That he should have proceeded to the demonstration of the laws of electrolysis, was the next most natural result. He showed that the amount of any compound decomposed by the electric current is exactly proportional to the whole quantity of electricity which has passed through the electrolyte. Different substances are variously refractory to dissolution under the influence of the electric current, but each one always acts in the same way and requires the same amount of current. Substances that are closely related to one another chemically, are also related to one another in the amount of electricity required to bring about decomposition of their various compounds. He showed, of course, that there are differences of electrical relationship that make the results produced in the decomposition of various compounds very different. Polarization, for instance, sets in to a much greater degree in the decomposition of some substances than of others. One consequence is that the resistance to the passage of the electric current differs markedly, and the opposing electromotive force will stop the current or hamper its effects in many cases, so that, until after actual experiment, the quantitative effect of the passage of the electric current through a solution cannot be determined.

Faraday's opinions as to the significance of electricity in the animal economy are very interesting because of his profound knowledge of electrical phenomena and their place in nature. It is all the more interesting because it is so simple, and most scientists would be apt to say that its very simplicity is a very taking argument for its truth. "As living creatures produce heat, and a heat certainly identical with that of our hearths, why should they not produce electricity also, and an electricity in like manner identical with that of our machines? Like heat, like chemical action, electricity is an implement of life, and nothing more."

While Faraday often occupied himself with subjects connected with matter and force that are likely to remain mysteries for long after his time, and often had thoughts to express with regard to the nature of atoms and of imponderable agents, whatever he had to say about these subjects was not vague and speculative, but, on the contrary, was concrete and usually of such a practical character as to add something new to our knowledge of them. Few men have ever succeeded in getting closer to the mysteries that underlie natural phenomena than Faraday; yet no one was ever less carried away into vague theoretic speculations with regard to them, nor tempted to think that because he knew much more than most other men with regard to complex natural problems, that therefore he knew enough to be able to solve the mysteries that existed all around him. He had none at all of what would ordinarily be called pride of intellect, but, on the contrary, had the humility of the true scientist. Knowing so much only made him realize more poignantly how much he was ignorant of. With regard to his speculations on matter and force and the imponderables, Helmholtz, the great German physicist, once summed up Faraday's contributions very succinctly in a way to show the practical nature of Faraday's intellect. He said:

"It is these things that Faraday in his mature works ever seeks to purify more and more from everything that is theoretical and is not the direct and simple expression of the fact. For instance, he contended against the action of forces at a distance, and the adoption of two electrical and two magnetic fluids, as well as all hypotheses contrary to the law of the conservation of force, which he early foresaw, though he misunderstood it in its scientific expression. And it is just in this direction that he exercised the most unmistakable influence, first of all, on the English physicist, and then on the physicists of all the world."

Inventors and promoters of useful inventions, frequently benefited by the advice of Faraday or by his general help. A remarkable instance of this was told by Mr. Cyrus W. Field. At the commencement of his great enterprise, when he wished to unite the Old and the New World by the telegraphic cable, he sought the advice of the great electrician, and Faraday told him that he doubted the possibility of getting a message across the Atlantic. Mr. Field saw that this fatal objection must be settled at once, and begged Faraday to make the necessary experiments, offering to pay him properly for his services. The philosopher, however, declined all remuneration, but worked away at the question, and presently reported to Mr. Field: "It can be done; but you will not get an instantaneous message." "How long will it take?" was the inquiry. "Oh! perhaps a second." "Well, that's quick enough for me," was the conclusion of the American; and the enterprise was proceeded with.

Faraday was far from being a mere laboratory student; he was much more even than a great teacher of physics. He was a magnificent popular lecturer, and did an incalculable amount to bring physics to the attention and the serious interest of his generation. A contemporary has described one of his lectures at the Royal Institution in such a way as to give us some idea, even at this distant date, of Faraday's power over his audience, of his own wonderful interest in the subject and his marvelous ability to communicate that interest to others. It was of the very nature of the man that he should not be cold and formal, for he was not a man of the head alone, but, above all, a man whose heart and affections were greatly developed, and he had powers of enthusiasm that placed him high among the artistic spirits of mankind. Our American poet, Stedman, once declared that the intellectual quality of the poet, the creator in the realm of thought, and of the scientist, the original worker in the domain of science, differed but little from one another, and must be considered as collateral expressions of the same form of intellectual genius. With this in mind, his contemporary's enthusiastic description of his lectures will not seem overdrawn.

"It was an irresistible eloquence, which compelled attention and insisted upon sympathy. It waked the young from their visions, and the old from their dreams. There was a gleaming in his eyes which no painter could copy, and which no poet could describe. Their radiance seemed to send a strange light into the very heart of his congregation; and when he spoke, it was felt that the stir of his voice and the fervor of his words could belong only to the owner of those kindling eyes. His thought was rapid, and made itself a way in new phrases, if it found none ready made, as the mountaineer cuts steps in the most hazardous ascent with his own axe. His enthusiasm sometimes carried him to the point of ecstasy."

Faraday's habit of testing opinions by experiment, and the frequent disillusions which he encountered with regard to things of which he thought he knew something definite, served to make him extremely careful as regards expressions of opinion. Some of his thoughts on this subject are worth while recalling because they remain perennially true, and anyone in any generation will find that, as his experience grows, he gets more and more into this Faraday mood of doubting his own opinion and listening with more readiness to that of others. As a rule, this is said not to be true of those who are in advancing years, but the greater minds among the older men do not get set in their ways. Flourens might have said that because of constant exercise the connective tissue in the brains of such men does not form to the same extent as in others, and does not make them case-hardened. As a consequence, they retain far on in years their sympathy for others' opinions and their openness of mind. Comparatively, they are so few, however, that this expression of Faraday's becomes a striking commentary on his large-mindedness.

"For proper self-education, it is necessary that a man examine himself, and that not carelessly either.... A first result of this habit of mind will be an internal conviction of ignorance in many things respecting which his neighbors are taught, and that his opinions and conclusions on such matters ought to be advanced with reservation. A mind so disciplined will be open to correction upon good grounds in all things, even in those it is best acquainted with, and should familiarize itself with the idea of such being the case."

Perhaps it is even more interesting, because more humanly sympathetic, to find that Faraday distrusted his opinions of people even more than his opinions of things, and that he himself tried to be very slow to take offence at what was said to him, and counselled greatest discretion to others in judging of the significance of supposed slights.

"Let me, as an old man who ought by this time to have profited by experience, say that when I was younger, I found I often misinterpreted the intentions of people, and found that they did not mean what at the time I supposed they meant; and further, that, as a general rule, it was better to be a little dull of apprehension when phrases seemed to imply pique and quick in perception, when, on the contrary, they seemed to imply kindly feeling. The real truth never fails ultimately to appear, and opposing parties, if wrong, are sooner convinced when replied to forbearingly than when overwhelmed."

Few lives have been happier than that of Faraday. He gave up the ordinary ambition of men to make what is called a successful career of money-making, and constantly guarded himself from slipping back, as so many do, to the ruin of their original purpose. He lived a long life in peace, occupied with work that he liked above all things, and surely serves as the best illustration of the maxim: "Blessed is the man who has found his work." Work is said to be one of the primal curses laid upon man; but if, when the Creator would ban it turns to blessing in the way that work has done, then may one well ask what will His blessings prove. Faraday even had what is rarer in life than happiness, the consciousness of his happiness. Usually it is so elusive that it escapes reflection. At the close of his career, when he wrote, in 1861, to the managers of the Royal Institution resigning most of his duties, he expressed this feeling very beautifully, and at the same time so simply and clearly as to make his letter of resignation a precious bit of literature.

"I entered the Royal Institution in March, 1813, nearly forty-nine years ago, and, with the exception of a comparatively short period, during which I was abroad on the continent with Sir H. Davy, I have been with you ever since. During that time I have been most happy in your kindness, and in the fostering care which the Royal Institution has bestowed upon me. Thank God, first, for all His gifts! I have next to thank you and your predecessors for the unswerving encouragement and support which you have given me during that period. My life has been a happy one, and all I desired. During its progress, I have tried to make a fitting return for it to the Royal Institution, and through it to science. But the progress of years (now amounting in number to three-score and ten) having brought forth, first, the period of development, and then that of maturity, has ultimately produced for me that of gentle decay. This has taken place in such a manner as to make the evening of life a blessing; for, while increasing physical weakness occurs, a full share of health, free from pain, is granted with it; and while memory and certain other faculties of the mind diminish, my good spirits and cheerfulness do not diminish with them."

For nearly five years after he had given up to a great degree his work at the Royal Institution, he faced death, not with the equanimity of the stoic, but with the peaceful happiness of the believer in Providence and a hereafter. Even the loss of his memory, dear as it must have been to a man who had spent all his life in storing it with the great facts of science, does not seem seriously to have disturbed him. He realized the necessity for patience, and took the lesson of its necessity to heart, so that there was no difficulty in it. Once when calling on his friend, the distinguished scientist, Barlow, who had for a lifetime almost worked beside him at the Royal Institution, but who was now suffering from paralysis, he said: "Barlow, you and I are waiting; that is what we have to do now; and we must try to do it patiently." When the full realization that his powers were leaving him first came to him, he wrote to his niece what he thought ought to be the feelings of the believer in Providence toward death, and his letter shows how thoroughly he had imbibed the great lessons of Christianity, and how much of consolation his faith was to him in this darkest hour before the dawn of that other life, in which he had as implicit confidence as in any of the great scientific principles that he had demonstrated by experiment. He wrote:

"I cannot think that death has, to the Christian, anything in it that should make it a rare, or other than a constant thought. Out of the thought of death comes the view of the life beyond the grave, as out of the view of sin (that true and real view which the Holy Spirit alone can give to man) comes the glorious Hope.... My worldly faculties are slipping away day by day. Happy is it for all of us, that the true good lies not in them. As they ebb, may they leave us as little children, trusting in the Father of Mercies and accepting His unspeakable gift." And when the dark shadow was creeping over him, he wrote to the Comte de Paris: "I bow before Him who is the Lord of all, and hope to be kept waiting patiently for His time and mode of releasing me, according to His divine word and the great and precious promises whereby His people are made partakers of the divine nature."

Probably the feature of the careers of Darwin and Spencer which are saddest for their adherents, and which made those who refused to be recognized as among their followers appreciate their one-sidedness, is the confession by both of them, that they had lost their interest in poetry and even in literature of all kinds, and toward the end of their lives particularly lost entirely their appreciation of things artistic. As might be expected from what we know of Faraday, this was not at all the case with him; but, on the contrary, down to the end of his life, he retained all his youthful admiration for the poets. His niece tells the story of hearing him often read poetry, and of how much he used to be affected by his favorite poems. In one of her letters she says:

"But of all things, I used to like to hear him read 'Childe Harold'; and never shall I forget the way in which he read the description of the storm on Lake Leman. He took great pleasure in Bryon, and

Coleridge's 'Hymn to Mont Blanc' delighted him. When anything touched his feelings as he read—and it happened not infrequently—he would show it not only in his voice, but by tears in his eyes also."

As a young man, he was so completely taken up with the scientific studies that he could not think that he would ever find time for the ordinary interests of life. Especially was this true with regard to the question of marriage. He felt that he would never marry, and he seems rather to have pitied those, the weakness of whose nature pushed them on to assume many duties in life and look for merely selfish happiness. It was as a very young man that he wrote:

"What is't that comes in false, deceitful guise, Making dull fools of those that 'fore were wise? 'Tis Love."

When the time came, however, he altered this opinion. Among the elders of the Church which he attended in London was a Mr. Barnard, a silversmith. Faraday occasionally spent an evening at his house, and incidentally met his daughter Sarah. He had not met her many times before his ideas as to what love might mean in life were completely changed, and not long after making her acquaintance he wrote her a letter, in which he recants and asks her to be more than a friend. His letter is rather interesting as love letters go.

"You know me as well or better than I do myself. You know my former prejudices and my present thoughts; you know my weaknesses, my vanity, my whole mind; you have converted me from one erroneous way; let me hope that you will attempt to correct what others are wrong.... Again and again I attempt to say what I feel, but I cannot. Let me, however, claim not to be the selfish being that wishes to bend his affections for his own sake only. In whatever way I can best minister to your happiness, either by assiduity or by absence, it shall be done. Do not injure me by withdrawing your friendship, or punish me for aiming to be more than a friend by making me less; and if you cannot grant me more, leave me what I possess but hear me."

In spite of the sincere feeling of this letter, the lady hesitated. For a time she left London, apparently in order to give herself a breathing spell from the ardor of his suit. In spite of his deep interest in science, Faraday followed her to the seacoast, and after they had wandered together for several days at Margate and Dover, where Shakespeare's Cliff Was an especial haunt of theirs, the lady relented. Faraday returned to London bubbling over with happiness. He was not quite thirty when they were married, and at the time his salary did not amount to more than a thousand dollars a year. It was distinctly not a marriage of reason.

Most of the happiness of his life came to him from his marriage. Many years afterward, he called it "An event which, more than any other, contributed to my happiness and healthful state of mind." With years, this feeling only deepened and strengthened. In the midst of his scientific triumphs, his first thought was always of her. When his attendance at scientific congresses took him away from her, his letters were frequent, and always expressive of his longing to be with her. One of his biographers has said "that doubtless at any time between their marriage and his final illness, he might have written to her as he did from Birmingham, at the time of the meeting of the British Association there."

"After all, there is no pleasure like the tranquil pleasure of home; and here, the moment I leave the table, I wish I were with you *in quiet*. Oh! what happiness is ours! My runs into the world in this way only serve to make me esteem that happiness the more."

Faraday had probably lost more illusions than most men, and came to the true appreciation of things as they are. In spite of his life-long study, he had no illusions with regard to the education of the intellect merely, or the possession of superior intellectual faculties as moral factors. His keen observation of men had made any such mistake as that impossible. On the other hand, he had often noted that the ignorant, or at least those lacking education, were very admirable in conduct and in principle, and so we have his suggestive testimony:

"I should be glad to think that high mental powers insured something like a high moral sense, but have often been grieved to see the contrary; as also, on the other hand, my spirit has been cheered by observing in some lowly and uninstructed creature such a healthful and honorable and dignified mind as made one in love with human nature. When that which is good mentally and morally meet in one being, that that being is more fitted to work out and manifest the glory of God in the creation, I fully admit."

Faraday's very definite expression of what he considers must be the position of the man of science with regard to a hereafter and the existence of God, is worth while recalling here, because it was such a modest yet forceful presentation of the attitude of mind that every thinking modern scientist must occupy in this matter, the attitude which all of Faraday's great fellow-workers in the domain of electricity also occupy. It is indeed the position that has been assumed by all the great scientists who bowed humbly to faith, though so many lesser lights have found this apparently impossible. At a lecture given in 1854 at the Royal Institution, Faraday said: "High as man is placed above the creatures around him, there is a higher and far more exalted position within his view; and the ways are infinite in which he occupies his thoughts about the fears, or hopes, or expectations of a future life. I believe that the truth of that future cannot be brought to his knowledge by any exertion of his mental powers, however exalted they may be; that it is made known to him by other teaching than his own, and is received through simple belief of the testimony given.... Yet even in earthly matters, I believe that 'the invisible things of Him from the creation of the world are clearly seen, being understood by the things that are made, even His eternal power and godhead'; and I have never seen anything incompatible between those things of man which can be known by the spirit of man which is within him, and those higher things concerning his future which he cannot know by that spirit."

Elsewhere he had said: "When I consider the multitude of associate forces which are diffused through nature; when I think of that calm and tranquil balancing of their energies which enables elements, most powerful in themselves, most destructive to the world's creatures and economy, to dwell associated together and be made subservient to the wants of creation, I rise from the contemplation more than ever impressed with the wisdom, the beneficence, and grandeur beyond our language to express, of the Great Disposer of all!"

Dr. Gladstone, in his Life of Faraday, which we have so often put into requisition, has given in one striking paragraph a description of the passing of Faraday, that in its simplicity is worthy of the great man whom it so well represents. It is so different from what is ordinarily supposed to be the attitude of the scientist towards death, that when by contrast we recall that Faraday is acknowledged to be the greatest experimental scientist of the nineteenth century, the man of his generation most honored by scientific societies at home and abroad—his honorary memberships numbered nearly one hundred—it must be considered as a very curious contradiction of what is the usual impression in this matter: "When his faculties were fading fast, he would sit long at the western window, watching the glories of the sunset; and one day, when his wife drew his attention to a beautiful rainbow that then spanned the sky, he looked beyond the falling shower and the many-colored arch and observed, 'He hath set His testimony in the heavens.' On August 25th, 1867, quietly, almost imperceptibly, came the release. There was a philosopher less on earth, and a saint more in heaven."

When we come to the end of the life of this greatest of experimentalists, the most striking remembrance is that of the supreme original genius of this great discoverer in electricity, whose work was such a stimulus to others, whose conclusions were to prove the basis for so much of the work of his contemporaries and his successors in electrical investigation, and whose place in the world of science is assured beside such men as Newton and Kepler and Harvey and the other great pioneers in science. There is no doubt at all, however, that our heartiest feelings are aroused by the picture of the wonderfully rounded existence of the great scientist, his pervasive humanity, his largeness of soul and sympathy, his understanding of men in their ways through his own complete knowledge of himself, that is so strikingly displayed. We feel sure that Faraday himself would have cared less for his fame as a great scientist than for the summary of his life which has been given us by his friend, Bence Jones, who said: "His was a life-long strife, to seek and say that which he thought was true and to do that which he thought was kind."

CHAPTER XI. CLERK MAXWELL.

Natural science in every department developed very wonderfully from its experimental side during the first half of the nineteenth century. Facts and observations accumulated to such an amount that, shortly after the middle of the century, there was felt the need of a great mathematical genius to bring the results of experiment into their proper places in the great body of applied and theoretic science. Nearly always such a demand meets with adequate response in its own due time. Clerk Maxwell came at this most opportune moment for science. No mathematical problem was too abstruse or difficult for him, and whatever he took up seriously he always illuminated, and usually solved its problems as completely as can be hoped for in the present state of scientific knowledge. It was particularly in electricity that his mathematical faculty proved of the greatest value, and that he found the abundant opportunities of which he knew so well how to take advantage.

Clerk Maxwell's theory of electricity, as developed in his classic treatise on "Electricity and Magnetism," is well called by Prof. Peter Guthrie Tait, "One of the most splendid monuments ever raised by the genius of a single individual." This book became the guide and companion of more physical scientists during the nineteenth century than perhaps any other written in that period. It was not alone in England or in English-speaking countries that it was accepted as an authority and constantly referred to, but everywhere throughout the world of science. Not to know it, was to argue that a man knew nothing of the profounder truths of electrical science and was only a seeker after superficial information. Clerk Maxwell was known and esteemed by all the great physical scientists of the world. His name is less widely known than that of most of the great discoverers in electricity, because mathematical achievement always has less popular attraction; but he deserves to be known by all who are interested in science, not only because of his magnificent contributions to mathematical electricity, but quite as much for qualities of heart and mind that stamp him as one of the very great men of the century so rapidly receding from us.

Clerk Maxwell, as he is usually called, because he was the representative of a younger branch of the well-known Scottish family of Clerk of Penicuik, was born in Edinburgh, June 13th, 1831. As with nearly every other person who reaches distinction in after-life, there are stories told of his precociousness which probably have more meaning in this case than in most others, since they exhibit real traits that were characteristic of the man. As a child, it is said that he was never satisfied until he had found out for himself everything that he could about anything that attracted his attention. He wanted to know where the streams of water came from, where and whence all the pipes ran, and the course of bell-wires and the like. His frequently repeated question was, "What's the go o' that." If an attempt were made to put him off with some indefinite answer, then he would insist, "But what's the particular go of it." This was probably the most prominent trait in his after-life. General explanations of phenomena that satisfied other men never satisfied him. He was a nature student from the beginning, and even as a boy he devised all sorts of ingenious mechanical contrivances. Pet animals were his special delight, but for experimental purposes always, and his selection of pets would probably have startled some people.

He received his early education at the Edinburgh Academy, and his university education at the University of Edinburgh, where he graduated in 1850. His liking for mathematics, which had already been very strongly exhibited, led him, at the age of nineteen, to go to Cambridge. Here, for a term or two, he was a student at Peterhouse, but afterwards found a more sympathetic place for his mathematical tastes at Trinity. He took his degree at Cambridge in 1854, though only with the rank of second wrangler, Routh being senior. In the more serious and more exacting examination for the Smith's Prize, he was declared equal with the senior wrangler. His mathematical talents had developed very early, and it is not surprising that the rest of his life should have been devoted mainly to the teaching of mathematics and in investigations connected with applied mathematics. It was not success at the university that determined his career, for he had shown his marvelous mathematical ability much earlier than that, and had given some astonishing examples of his power to treat complex scientific problems in mathematical journals.

Indeed, his original contributions to the higher mathematics began before he was fifteen years of age. He was a striking example of the fact that a great genius usually finds his work very early in life, and usually accomplishes something significant in it, at once the harbinger and the token of the future, before he is twenty-five. While Clerk Maxwell was at the Edinburgh Academy, Professor J. D. B. Forbes, in 1836, communicated to the Royal Society of Edinburgh a short paper by his youthful student on "A Mechanical Method of Tracing Oval Curves" (Cartesian Ovals).

In spite of the prejudice that exists with regard to precocious genius and the distinct feeling that it is not likely to prove an enduring quality, Clerk Maxwell continued to do excellent original work all through his teens. When he was but eighteen, he contributed two important papers to the transactions of the Royal Society of Edinburgh. One of these was on "The Theory of Rolling Curves," and the other on "The Equilibrium of Elastic Solids." These are now remembered, not only because of Clerk Maxwell's subsequent distinguished career, but because of their distinct value as contributions to science. Both of them demonstrate not only his ability to work out subtle mathematical problems at this very early age, but show the possession by him of a power of investigation for original work that stamps them as well worthy of consideration in themselves, quite apart from the repute of their author or the successful accomplishments of his subsequent life.

With regard to one of those Edinburgh papers of Clerk Maxwell's eighteenth year, Prof. Guthrie Tait said "that in it he laid the foundation of one of the singular discoveries of his later life, the temporary double refraction produced in viscous liquid by sheering stress." After his magnificent mathematical training at Cambridge, it is not surprising that this academic career of great original work should be continued by contributions to science of ever-increasing importance. Immediately after his graduation, he read to the

Cambridge Philosophical Society one of the few purely mathematical papers that he ever published. This had for its title, "On the Transformation of Surfaces by Bending." Expert mathematicians who read the paper, realized at once that there was a new genius in the field of mathematics. During the same year, the young Scotch mathematician took the first step in that series of electrical investigations which was to occupy so much of his attention in after-life, and which was to prove the source of his greatest inspirations. This consisted of the publication of an elaborate paper on Faraday's "lines of force."

While we think of Maxwell as a mathematical physicist, it must not be forgotten that he was also one of the leading experimental scientists of that great epoch, the nineteenth century. Only a man who was himself a great experimenter could have properly appreciated and developed, from the mathematical standpoint, the works of such men as Cavendish and Faraday. From his early years, Maxwell displayed a distinct fondness for experimentation, and this even extended to experiments upon himself. In many ways this trait of his would remind us of Johann Müller, the great father of modern German medicine. Like Müller, there was danger also of Maxwell's experiments on himself getting him into trouble. For instance, at one time his love of experiment led him to try sleeping in the evening and getting up to work at midnight, so as to have the long, silent hours of the night to himself. In the sketch of his life by Dr. Garnett, a letter from one of his friends is quoted with regard to this nocturnal habit, which is amusing as well as interesting. The friend wrote:

"From 2 to 2:30 a. m. he took exercise by running along the upper corridor, down the stairs, along the lower corridor, then up the stairs, and so on until the inhabitants of the rooms along his track got up and laid *perdus* behind their sporting doors, to have shots at him with boots, hair-brushes, etc., as he passed." His love of fun, his sharp wit, his extensive knowledge, and, above all, his complete unselfishness, rendered him a universal favorite, in spite of the temporary inconveniences which his experiments may have occasionally caused to his fellow-students.

In 1857, Clerk Maxwell received the Adams Prize for his essay on "The Stability of the Motion of Saturn's Rings." He shows very clearly that these annular appendages consist of a large number of small masses. This work would seem to be very distant from anything that Maxwell had attempted before, and would indeed seem to the superficial observer, at least, to be quite out of his sphere. It was the mathematics of it that attracted him, and the fact that the problem was difficult, indeed, one of the most difficult at that time before astronomers, only added zest to his resolve to fathom it. All his life, mathematics continued to be his favorite form of work, and his power to express the most complex physical phenomena in mathematical formulæ gave him a reputation throughout Europe unsurpassed by anyone of his generation. The more a problem seemed incapable of direct statement in mathematical terms, provided it represented a great occurrence in nature, the more Maxwell was attracted to it; and the training of these early years in thus setting mathematics to the solution of physical relations, was to serve him in good stead when he came to try his hand at demonstrating the meaning of electricity in mathematical terms.

Just before this, in 1856, Maxwell, though only twenty-five years of age, was offered the chair of natural history, which included most of the physical sciences, at Marischal College, Aberdeen. With the attention that his mathematical papers attracted, it is not surprising that after four years of teaching experience he was invited to King's College, London. He held his new position for eight years, and then his health required him to retire to his estate in Kirkcudbrightshire. After three years of retirement, his English Alma Mater demanded his services, and the temptation to get back to an academic career was so great that he could not resist it. He became, in 1871, Professor of experimental physics at Cambridge. To him, more than to anyone else, is due the magnificent development of the physical sciences which took place at Cambridge during the last quarter of the nineteenth century. Unfortunately, he was not destined to live to enjoy the fruits of his labor in organizing the scientific side of the university, but it was under his direction that the plans of the Cavendish Laboratory were prepared, and he superintended every step of the progress of the building. It was under his careful management, too, that the purchase of the very valuable collection of apparatus, with which it was equipped by the Duke of Devonshire, was made, and Maxwell's work here counts for much in the history of English science.

He died in 1879, when only forty-eight years of age, but he had deeply impressed himself upon the science of the nineteenth century. For quite one-half of his scant half-century span of life he had occupied a prominent place in England, and after the age of thirty-five had come to be generally recognized as one of the leading physical scientists of the world. His career is, as we have said, a striking illustration of how early in life a man's real work is likely to come to him, and how little success in original investigation is dependent on that development of mind which is supposed to be due only to long years of application to a particular branch of study. Manifestly it is the original genius that counts for most, and not any training that it receives, except such as comes from its own maturing powers. Environment, if unfavorable, does not hamper it much, nor keep it from reaching the proper terminus of its destiny; and poor health only serves to prevent the exercise of its full powers, but does not eclipse the manifestation of its capacity.

Clerk Maxwell's important contribution to science was the demonstration that electro-magnetic effects travel through space in the form of transverse waves similar to those of light and having the same velocity. We have become so familiar with the ideas contained in this explanation, that they seem almost obvious now. They came, however, as a great surprise to Clerk Maxwell's generation, and at first seemed to be merely a theoretic expression of a mathematical formula. Not long afterwards, however, Maxwell's explanation was corroborated by Hertz, who showed that these waves were propagated just as waves of light are, and that they exhibit the phenomena of reflection, refraction and polarization. Hertz went on from his demonstration of the actuality of Maxwell's mathematical theory to the demonstration of further electrical waves. These Hertzian waves, as they were called, were a startling discovery, but remained only a scientific curiosity until they were taken advantage of for wireless telegraphy, when a new era of applied electrical science began.

How his success in this was accomplished will be best understood from Prof. Guthrie Tait's account of Maxwell's devotion to electricity as a life-work. He says:

"But the great work of his life was devoted to electricity. He began by reading with the most profound admiration and attention the whole of Faraday's extraordinary self-revelations, and proceeded to translate the ideas of that master into the succinct and expressive notation of the mathematicians. A considerable part of this translation was accomplished during his career as an undergraduate in Cambridge. The writer had the opportunity of perusing the MS. on Faraday's lines of force, in a form little different from the final one, a year before Maxwell took his degree. His great object, as it was also the great object of Faraday, was to over-turn the idea of action at a distance. The splendid researches of Poisson and Gauss had shown how to reduce all the phenomena of statical electricity to mere attractions and repulsions exerted at a distance by particles of an imponderable on one another. Sir W. Thomson had, in 1846, shown that a totally different assumption, based upon other analogies, led (by its own special mathematical methods) to precisely the same results. He treated the resultant electric force at any point as an analogous flux of heat from the sources distributed, in the same manner as the supposed electric particles. This paper of Thomson's, whose ideas Maxwell afterwards developed in an extraordinary manner, seems to have given the first hint that there are at least two perfectly distinct methods of arriving at the known formulæ of statical electricity. The step to magnetic phenomena was comparatively simple; but it was otherwise as regards electromagnetic phenomena, where current electricity is essentially involved. An exceedingly ingenious, but highly artificial, theory had been devised by Weber, which was found capable of explaining all the phenomena investigated by Ampère as well as the induction currents of Faraday. But this was based upon the assumption of a distance-action between electric particles, whose intensity depended upon their relative motion as well as on their position. This was, of course, more repugnant to Maxwell's mind than the statical distance-action developed by Poisson. The first paper of Maxwell's in which an attempt at an admissible physical theory of electromagnetism was made, was communicated to the Royal Society in 1867. But the theory in a fully developed form, first appeared in his great treatise on Electricity and Magnetism (1873). Availing himself of the admirable generalized coördinate system of Lagrange, Maxwell has shown how to reduce all electric and magnetic phenomena to stresses and motions of a material medium, and as one preliminary, but excessively severe, test of the truth of this theory has shown that, if the electromagnetic medium be that which is required for the explanation of the phenomena of light, the velocity of light in vacuo should be numerically the same as the ratio of the electromagnetic and electrostatic units. We do not as yet certainly know either of these quantities very exactly, but the mean values of the best determination of each separately agree with one another more closely than do the various values of either. There seems to be no longer any possibility of doubt that Maxwell has taken the first grand step towards the discovery of the true nature of electrical phenomena. Had he done nothing but this, his fame would have been secure for all time. But, striking as it is, this forms only one small part of the contents of his truly marvelous work."

Maxwell's prediction as to the propagation of electric waves has received its full confirmation, as we have said, in the brilliant experiments of Hertz, and in the subsequent application of the Hertzian waves to wireless telegraphy in our own time. It was not by mere chance that this development of Maxwell's thinking came. Hertz himself declared, in the introduction to his collected papers, that he owed the suggestion of his work to Faraday and Maxwell, and above all to Maxwell's speculations as to the nature of electricity and its relations to light. Hertz said:

"The hypothesis that light is an electric phenomenon is thus made highly probable. To give a strict proof of this hypothesis would logically require experiments upon light itself. There is an obvious comparison between the experiments and the theory, in connection with which they were really undertaken. Since 1861, science has been in possession of a theory which Maxwell constructed upon Faraday's views, and which we therefore call the Faraday-Maxwell theory. This theory affirms the occurrence of the class of phenomena here discovered, just as positively as the remaining electric theories are compelled to deny it. From the outset, Maxwell's theory excelled all others in its elaboration and in the abundance of relations between the various phenomena which it included."

How much Maxwell's work was appreciated across the channel, may be realized from what Poincaré said: "So sure did the results of his (Maxwell's) theory appear as worked out for the deepest problems, that a feeling of distrust and suspicion is likely to be mingled with our admiration for his magnificent work. It is only after prolonged study and at the cost of many efforts that this feeling is dissipated."

Maxwell's explanation of electricity is that it is a strain or stress in the ether, that it is a condition or mode, and not a substance. One distinguished foreign contemporary who had read Maxwell's books with the greatest interest, declared that he could not be quite satisfied, since nowhere did he find what a charge of electricity is, though he seemed to find satisfactory information with regard to everything else. Maxwell realized, however, the limitations of his speculation very well, and hesitated, above all, to bind his mathematical conclusions to statements that might prove eventually only surmises founded on insufficient information from the standpoint of observation. Even when he gave his explanation, he did not insist on it as absolute, but, as pointed out by Poincaré, discussed it only as a possibility. The French scientist said: "Maxwell does not give a mechanical explanation of electricity and magnetism; he is only concerned to show that such an explanation is possible."

Maxwell thoroughly believed in having a hobby as well as his regular work, and during the time while he was devoting himself to the mathematical explanation of electricity he turned for recreation to certain problems in physics, in physiology and psychology, relating to color. He worked almost as great a revolution in our knowledge of color-vision as in any other subject that he took up. Principal Garnett has condensed so well what Clerk Maxwell accomplished in the matter of color-vision, in his sketch of him in "The Heroes of Science," that I prefer to quote his explanation. He says:

"It has been stated that Thomas Young propounded a theory of color-vision which assumes that there exists three separate color sensations, corresponding to red, green and violet, each having its own special organs, the excitement of which causes the perception of the corresponding color, other colors being due to the excitement of two or more of these simple sensations in different proportions. Maxwell adopted blue instead of violet for the third sensation, and showed that, if a particular red, green, and blue were selected and placed at the angular points of an equilateral triangle, the colors formed by mixing them being arranged as in Young's diagram, all the shades of the spectrum would be ranged along the sides of this triangle, the center being neutral grey. For the mixing of colored lights, he at first employed the color top; but instead of painting circles with colored sectors, the angles of which could not be changed, he used circular discs of colored paper slit along one radius. Any number of such discs can be combined so that each shows a sector at the top, and the angle of each sector can be varied at will by sliding the corresponding disc between the others. Maxwell used discs of two different sizes, the small discs being placed above the larger on the same pivot, so that one set forked a central circle and the other set a ring surrounding it. He found that, with discs of five different colors, of which one might be white and another black, it was always possible to combine them so that the inner circle and the outer ring exactly matched. From this he showed that there could be only three conditions to be satisfied in the eye, for two conditions were necessitated by the nature of the top, since the smaller sectors must exactly fill the circle and so must the larger. Maxwell's experiments, therefore, confirmed, in general, Young's theory. They showed, however, that the relative delicacy of the several color sensations is different in different eyes, for the arrangement which produced an exact match in the case of one observer, had to be modified for another; but this difference of delicacy proved to be very conspicuous in color-blind persons, for in most of the cases of color-blindness examined by Maxwell the red sensation was completely absent, so that only two conditions were required by color-blind eyes, and a match could therefore always be made in such cases with four discs only. Holmgren has since discovered cases of color-blindness in which the violet sensation is absent. He agrees with Young in making the third sensation correspond to violet rather than blue. Maxwell explained the fact that persons color-blind to the red divide colors into blues and yellows, by the consideration that, although yellow is a complex sensation corresponding to a mixture of red and green, yet in nature, yellow tints are so much brighter than greens, that they excite the green sensation more than green objects themselves can do; and hence greens and yellows are called yellow by such color-blind persons, though their perception of yellow is really the same as perception of green by normal eyes. Later on, by a combination of adjustable slits, prisms, and lenses arranged in a 'color box,' Maxwell succeeded in mixing, in any desired proportions, the light from any three portions of the spectrum, so that he could deal with pure spectral colors instead of the complex combinations of differently colored lights afforded by colored papers. From these experiments, it appears that no ray of the solar spectrum can affect one color sensation alone, so that there are no colors in nature so pure as to correspond to the pure simple sensations, and the colors occupying the angular points of Maxwell's diagram affect all three color sensations, though they influence two of them to a much smaller extent than the third. A particular color in the spectrum corresponds to light which, according to the undulatory theory, physically consists of waves, all of the same period; but it may affect all three of the color sensations of a normal eye, though in different proportions. Thus yellow-light of a given wave-length affects the red and green sensations considerably and the blue (or violet) slightly, and the same effect may be produced by various mixtures of red or orange and green."

For his researches on the perception of color, the Royal Society awarded Clerk Maxwell the Rumford Medal in 1860.

Besides this more or less theoretic work, however, Maxwell made some interesting and important discoveries and inventions in optics. For instance, he noted the great differences that exist in the eyes of dark and fair complexions to different colors when the light falls upon the center of the yellow spot, the so-called fovea centralis, or central pit of the retina. His researches with regard to this led him to the discovery that this portion of the retina is largely lacking in sensibility to blue light. He was able to demonstrate this by his experiment of looking through a glass vessel containing a solution of chrome alum, when the central portion of the field of vision appears of a light red color for the first second or two. He was also the inventor of an ingenious optical apparatus, a real image stereoscope. A still more important discovery was that of the double refraction which is produced for the time in viscous liquids when they are stirred and their motion is not as yet stopped. Maxwell showed that Canada balsam, for instance, when stirred, acquired a distinct power of double refraction, which it retained so long as the stress in the fluid produced by stirring remained.

Other departments of physics were not neglected. For instance, one of his greatest investigations was that on the kinetic theory of gases. Geniuses had been working before him on this line, for, as pointed out by Professor Tait, this theory owed its origin to Daniel Bernoulli, the greatest mathematician of the eighteenth century, and had been developed by the successful labors of Herapath, Joule and, above all, of Clausius. The work of these men put the general accuracy of the theory beyond all doubt and led to its very general acceptance, yet the details of it needed to be elaborated before it could become definitely scientific. Its greatest developments are due to Maxwell, and in this field Maxwell appeared as an experimenter on the laws of gaseous friction as well as a mathematician. His work with regard to color had showed his ingenuity as an experimentalist, and this is still further illustrated by his carefully arranged experiments on gases. Indeed, his work in this line makes it very clear that nothing was too difficult for him, and that anything that he turned his hand to in the field of science he was sure to accomplish with eminent success.

It was not only his scientific monographs, however, that indicate how great a scientist Clerk Maxwell was, but his text-books, even those of more or less elementary character, which he wrote bring out this same idea. He wrote, for instance, an admirable text-book on the theory of heat, which went through many editions. Students of the subject, even those who were not far advanced, found it clear and easier of study than many a less exhaustive work. He also wrote an elementary treatise on matter and motion, which has gone through several editions. One might think that so small a work would scarcely interest him enough to tempt him to put forth his powers at their best, and that at most it would be a conventional condensation of previous knowledge. Prof. Tait, who surely must be taken as a good judge in the matter, says that "even this, like his other and larger works, is full of valuable material worthy of the most attentive perusal not of students alone, but of the very foremost scientific men."

One of the characteristic traits of Maxwell was his desire to impart information to others. This extended not only to his academic relations, but, above all, to the working classes, who might have few opportunities for the obtaining of the information that was so interesting with regard to natural subjects. Everywhere that he held an academic post in his life, he gave lectures to the workmen. He was an extremely interesting talker, and one of his friends said of him: "I do believe there is not a single subject on which he cannot talk, and talk well, too, displaying always the most curious and out-of-the-way information." One of his private tutors said of him: "It is not possible for Maxwell to think incorrectly on physical subjects." It is easy to understand, then, how much his lectures to the working people at Aberdeen, at Edinburgh, and at Kings College, London, as well as at Cambridge, meant for them. If men like Maxwell would take up the popularization of science generally, then there would be much less opprobrium attached to the expression popular science than there has been only too often in the past, and is even at present.

Just as Maxwell set himself to the solution of the most difficult problems in physics, so he did not hesitate to give himself also to the discussion of problems in ethics. Here his power of penetration, the rigid logic of his mind, and his power to follow out conclusions to their ultimate significance, were quite as manifest as any scientific writing. It is almost the rule to find that scientists either ignore the great problems of man's place in nature and his destiny, or treat them very superficially. Agnosticism had become the fad of the moment, and was just beginning to make itself felt as a fashion in thinking when Clerk Maxwell was doing his great work. Maxwell was not an agnostic in science, and because he could not solve all the problems that came to him with regard to electricity and the constitution of matter, this did not keep him from setting himself to the task of seeing what should be his thoughts with regard to these subjects. He had none of the agnostic's feelings with regard to them, that since we cannot know all about them definitely and absolutely, therefore it is not worth while studying them at all. Had Maxwell been tempted to any such line of thought, we would have missed some of the most helpful scientific speculations and suggestions that have ever been made.

No one knew better than Maxwell, that his speculations on matter and electricity were theories, and that what he was offering to science were not definite explanations, but possible hypotheses. He has emphasized this himself over and over again. This inability of the human intellect at the present moment to solve all the questions that its inquiring spirit can evoke, did not keep him from investigating and following up his investigations by mathematical deductions and mechanical suggestions just as far as possible. He had the same attitude of mind toward the great problems of man's relation to his fellow-man, to the universe, and to a hereafter. While he felt that he could not solve the problems entirely, he felt also that his reasoning was quite sufficient to enable him to get a little nearer to the heart mystery of them and to understand something of their significance. In his later years, the question of the existence of pain and suffering in the world had, because of Darwin's attitude towards them and his declaration that since he was unable to understand them they carried him away from the thought of a beneficent Creator, attracted much attention. We have an essay of Clerk Maxwell's, then, on "Aspects of Pain," in which he discusses particularly pain as discipline. It is, of course, the old story, that men rise on stepping-stones of their dead selves, and that the successive deaths of self represent a triumphant progress, but it comes with a new vigor from this great scientist. We all know that it is the man who has suffered who is able to do things, and we are all well aware that the man who has lived in comfort all his life is almost sure to be lacking in character when a great crisis comes upon him. Indeed, as Clerk Maxwell re-states it, this is such a commonplace that one wonders why the problem of pain should have seemed so hard to understand.

There is an essay of his, also, on "Science and Free Will," which seems to deserve special notice. He has no illusions with regard to determinism. He is perfectly sure that he is free and that the great majority of men around him do or do not things as they choose. He points out that science makes for determinism only if one takes a very narrow view of it. Free will is not only compatible with scientific thinking, but it represents what would be expected as a culmination of the significance of life. In a word, Clerk Maxwell wrote as suggestively with regard to the great problems of human life as with regard to the physical nature around him that claimed so much of his interest. He was a true natural philosopher, and his interests were not limited merely to the lower orders of beings.

Because of the supreme power of Clerk Maxwell's mind to seek out the very heart of difficulties, the conclusions which he reached with regard to the existence of matter and the causes for the ultimate qualities which it exhibits, have an enduring interest. Mathematics is sometimes said to lead minds into scepticism. Cardinal Newman even thought that the mathematical cast of mind was the farthest removed from that which might be expected to accept things confidently on faith. Clerk Maxwell's intellect was eminently mathematical; yet, far from sending him over into the camp of the agnostics, his tendency to get at the ultimate reasons for things seemed almost to push him to conclusions with regard to the origin of matter, and especially its ultimate constituents, not ordinarily supposed to be scientific. A passage like the following, for instance, which may be found in his book on "The Theory of Heat," London, 1872, page 312, brings out this tendency very well:

"But if we suppose the molecules to be made at all, or if we suppose them to consist of something previously made, why should we expect any irregularity to exist among them? If they are, as we believe, the only material things which still remain in the precise condition in which they first began to exist, why should we not rather look for some indication of that spirit of order, our scientific confidence in which is never shaken by the difficulty which we experience in tracing it in the complex arrangements of visible things, and of which our moral estimation is shown in all our attempts to think and speak the truth, and to ascertain the exact principles of distributive justice?"

The argument from design for creation is often said in our day to have lost its weight. For Clerk Maxwell, however, this was evidently not the case. On the contrary, he seemed to find in the detailed knowledge of the ultimate constituents of matter which had come in recent years, additional proofs of the great design which permeates nature. He had come to the conclusion that not only were the groups of atoms which make up living things so ordered as to produce definite results, because there was a great purpose and, above all, a great Designer behind nature, but he also reached the position that the separate atoms of matter were so ordered with regard to one another, and in that ordering were so closely related to corresponding qualities in higher beings, that only the presence of a great design in nature could possibly account for all these wonderful attributes, which were to be found even in the smallest portions of matter. He said in his article on the atom, in the ninth edition of the Encyclopedia Britannica:

"What I thought of was not so much that uniformity of result which is due to uniformity in the process of formation, as a uniformity intended and accomplished by the same wisdom and power of which uniformity, accuracy, symmetry, consistency, and continuity of plan are as important attributes as the contrivance of the special utility of each individual thing."

Here is the old argument for the existence of God, from the design exhibited in the universe, rehabilitated by its application to the minutest portions of matter, whose qualities demand such an explanation quite as much as the highest adaptations of nature.

Perhaps the most striking expression of all with regard to the atoms that Clerk Maxwell permitted himself, is that in which he finds the type of what is best in man, in every minute portion of the universe, planted there by the Creator just as surely as they are in His highest beings, because they represent the most precious qualities of His own nature as they are reflected in the creation that He called into existence.

"They (the atoms) continue this day as they were created, perfect in number and measure and weight, and from the ineffaceable characters impressed on them we may learn that those aspirations, after accuracy in measurement, truth in statement, and justice in action, which we reckon among our noblest attributes as men, are ours because they are essential constituents of the image of Him Who in the beginning created not only the heaven and the earth, but the materials of which heaven and earth consist."

A very interesting side of Maxwell's life is that which shows his continued interest in literature, and even his occasional dippings into poetry. Though he reached distinction in mathematics and physics so early in his career, he yet found time to indulge a liking for the classics, and we even find some rather good translations of Horace's odes from his pen. The translation of a part of the Ajax of Sophocles from the Greek is a striking testimony to the breadth of Maxwell's intellectual interests. All during life, however, he permitted himself occasionally the luxury of fitting words into verse forms, and sometimes with a success that deserves much more than passing interest. It is very probable that the following verses, for instance, which are the first and last stanzas of a poem on the formula for being happy in life and were meant to be sung (or at least so he would hint) to the tune of "Il segreto per esser felice," will strike many a sympathetic chord in the modern time.

There are some folks that say They have found out a way To be healthy and wealthy and wise:— "Let your thoughts be but few, Do as other folks do, And never be caught by surprise. Let your motto be follow the fashion, But let other people alone; Do not love them nor hate them nor care for their fate, But keep a lookout for your own. Then what though the world may run riot, Still playing at catch who catch can, You may just eat your dinner in quiet And live like a sensible man."
In Nature I read quite a different creed, There everything lives in the rest; Each feels the same force As it moves in its course, And all by one blessing are blest. The end that we live for is single, But we labor not therefor alone; For together we feel how by wheel within wheel We are helped by a force not our own. So we flee not the world and its dangers, For He that has made it is wise; He knows we are pilgrims and strangers, And He will enlighten our eyes.

There probably was not a more nicely logical or more accurately reasoning intellect among all our nineteenth century scientists than that of the great mathematical electrician. He had none of the one-sidedness of the merely experimental scientist, nor, on the other hand, the narrowness of the exclusively speculative philosopher. With a power of analysis that was seldom equaled during the century, he had a power of synthesis that probably surpassed any of his contemporaries in any part of Europe. His ideas with regard to matter and its ultimate constitution are most suggestive. His suggestion of a strain in the ether as an explanation of electricity, thus enabling scientists to get away from the curious theories of the foretime which had required them to accept "action at a distance," that is, without any connecting medium, shows his power of following out abstruse ideas to definite practical conclusions. His religious life, then, will be a surprise to those who think that science leads men away from religion.

In the life of Clerk Maxwell, written by Campbell and Garnett, there is a passage from his friend and sometime pastor, Guillemard, in which the details of his religious life are given so fully as scarcely to require any further gleaning of information in this regard.

"He was a constant, regular attendant at church, and seldom, if ever, failed to join in our monthly late celebration of Holy Communion, and he was a generous contributor to all our parish charitable institutions. But his illness drew out the whole heart and soul and spirit of the man; his firm and undoubting faith in the Incarnation and all its results; in the full sufficing of atonement; in the works of the Holy Spirit. He had gauged and fathomed all the schemes and systems of philosophy, and had found them utterly empty and unsatisfying—'unworkable' was his own word about them—and he turned with simple faith to the Gospel of the Saviour."

His faith was not disturbed at the near approach of death, but, on the contrary, seemed strengthened. His biographers tell the story of some of the expressions used to his friends during these last days, which furnish manifest proof of this. Some of these passages are so characteristic and so striking that they deserve to be in the note-book of those to whom the modern idea that science is opposed to religion or faith may sometimes have been a source of worry, or at least an occasion for argument. Here is a typical one of these passages:

"Mr. Colin Mackenzie has repeated to us two sayings of his during those last days, which may be repeated here: 'Old chap, I have read up many queer religions; there is nothing like the old thing, after all; and I have looked into most philosophical systems, and I have seen that none will work without a God.'"

It must not be imagined, because Clerk Maxwell was a deeply religious man, that, therefore, he was frigid or formal or extremely serious, or inclined to be puritanic with regard to the pleasures of life, or a fanatic in the matter of taking all the good-natured fun there might be in anything that turned up. He was far from over-serious, or what has been called, though not quite properly, ascetic; but, on the contrary, was often, indeed usually, the soul of the party with which he was at the moment. He had none at all of the self-centered interest of the narrow-minded, but had many friends, and was liked by all his acquaintances. His friends were enthusiastic about his kindness of heart and the thorough congeniality of his disposition. On this point, the sketch of him in the National Dictionary of Biography gives a charming picture:

"As a man, Maxwell was loved and honored by all who knew him; to his pupils, he was the kindest and most sympathetic of teachers; to his friends, he was the most charming of companions, brimful of fun, the life and soul of a Red Lion dinner at the British Association meetings; but in due season brave and thoughtful, with keen interest in problems that lay outside the domain of his own work, and throughout his life a stern foe to all that was superficial or untrue. On religious questions, his beliefs were strong and deeply rooted."

It may be added to this, that his religion had nothing of the merely formal about it, nor was it perfunctory. It entered into most of the details of his life, and the fact that, every day as the head of the house he led evening prayers for the family, was only a token of the deep hold which religion had upon his life. When his last illness came, though he knew that his end was not far off, and at his age sometimes the approach of death hampers religious faith because it does seem that longer life might be afforded to one who has been so faithful in his realization of the obligations of life, Clerk Maxwell's piety increased rather than diminished. A favorite expression of his during his last days was the verselet from Richard Baxter, which one would be apt to think of as frequently repeated by some feminine devotee rather than by the greatest mathematical scientist of the nineteenth century:

"Lord, it belongs not to my care, Whether I die or live; To love and serve Thee is my share, And that Thy grace must give."

A friend who knew him intimately says: "In private life, Clerk Maxwell was one of the most lovable of men, a sincere and unostentatious Christian. Though perfectly free from any trace of envy or ill-will, he yet showed on fit occasions his contempt for that pseudo-science which seeks for the applause of the ignorant by professing to reduce the whole system of the universe to a fortuitous sequence of uncaused events."

In these phases of his intellectual life, the greatest of the mathematical electricians of the nineteenth century deserves to be taken as the type of the man of science, rather than the many mediocre intelligences whose minds were not large enough apparently for the two sets of truths — those of the moral as well as of the physical order.

Few men lived to witness so many remarkable discoveries in science and so many applications of the same to the welfare of the race as did the man whose name stands at the head of this chapter. When William Thomson, the future Lord Kelvin, first saw the light of day, the voltaic pile was in a rudimentary and inefficient form. It is true that water had been decomposed by the current from a pile in 1800, that the magnetic effect of the current had been discovered in 1820, and the possibility of a practical form of an electric telegraph suggested in the same year; but Ohm's law was still one of nature's secrets, electromagnetic induction was undiscovered, and the doctrine of energy but ill understood. Light, electricity and magnetism were regarded as distinct forces, and heat was thought to be a material substance, to which the name caloric was assigned. What Young, Fresnel and Ampère were in the early years of the nineteenth century; what Faraday, Regnault and Joseph Henry were some time later, Kelvin became in the 'fifties, a leader in the intellectual and scientific life of the time, a leader destined to extend the frontiers of knowledge, to establish an accurate system of electrical measurement, and to enrich the world with instruments of marvelous ingenuity and precision.

William Thomson, born in Belfast in 1824, received his early training in the Royal Academic Institute of that city. When eight years of age, he left his native land, exchanging the shores of Antrim for the banks of the Clyde. His father, James Thomson, a mathematician of note, having been appointed to the chair of mathematics in the University of Glasgow (founded in 1451), proceeded early in the summer of 1832 to the commercial metropolis of Scotland, accompanied by his two sons William and James, both of whom were destined to add lustre to the family name.

After a period of preparatory study, the two brothers, who were ten and eleven years of age, respectively, matriculated at the university. With the iron-clad regulations that govern admission to American colleges and universities, these boys would at best have been admitted to one of our high schools, and kept there until they reached the maturity required by the age limit. By the time young William attained that limit, he had already finished his work at the university, and captured the first prizes in mathematics, astronomy and natural philosophy. He was then only sixteen years of age, small of stature, but a giant in intellect; brilliant, versatile, and with a passion for work. It was his good fortune, also, to come under the influence of a great teacher, in the person of Prof. Nichol. "I have to thank what I heard in the natural philosophy class," he said in 1903, "for all I did in connection with submarine cables. The knowledge of Fourier was my start in the theory of signaling through submarine cables, which occupied a large part of my after-life. The inspiring character of Dr. Nichol's personality and his bright enthusiasm live still in my mental picture of those old days."

Having heard Fourier's treatise on the mathematical theory of heat spoken of one day as a remarkable and inspiring work, young Thomson astonished the Professor when, at the end of the lecture, he addressed Dr. Nichol with the query, "Do you think that I could read it?" To which the Professor smilingly replied: "Well, the mathematical part is very difficult." Many a student would have left Fourier alone for the nonce, after listening to a statement so little calculated to excite courage or awaken interest: but Thomson was not an ordinary student; and, however forbidding the answer which he received, he was determined all the same to handle the volume and seek its inspiration. Without delay, he got the book from the university library, and grew so delighted with the new ideas of the French mathematician about sine-expansions and cosine-expansions, that in the space of two weeks he had "turned over all the pages" of the book, as he modestly put it.

In the summer of 1840, he accompanied his father and his brother on a tour through Germany, partly to see the country and partly also, to acquire a practical knowledge of the language. In both these objects, he was somewhat hindered by his fondness for mathematical studies, which led him to include in his impedimenta for the trip a copy of Fourier's *Théorie analytique de la Chaleur*. Most students out on a summer's vacation, especially in foreign parts, would doubtless have preferred to give their minds rest and congenial distraction rather than keep on reading and pondering over abstract mathematical concepts. Our young tourist, on the other hand, seems to have thought of little else than of Fourier's "mathematical poem," as Clerk Maxwell called the work, a "poem" that continued to have a charm for him all through life. It is a noteworthy fact that Thomson continually returned to the ideas and methods of this suggestive treatise on the flow of heat, and that he applied them with great success to problems in thermal conductivity, in electricity and in submarine telegraphy.

Shortly after returning home, Thomson was sent to the University of Cambridge, where he entered St. Peter's College, commonly called Peterhouse, one of the oldest colleges of the university, its foundation dating back to the year 1284. Though he, no doubt, followed in a general way the directions given him by William Hopkins, "the best of private tutors," and kept in view the requirements of the honors examination, called the "Mathematical Tripos," for which he intended to present himself at the end of his course, he found his studies somewhat routinal and uninspiring. Original work was more to his taste than conventional subjects; his tutor, however, thought mainly of placing this brilliant pupil at the head of the wranglers, and hailing him the senior wrangler of the year, for which purpose, the beaten track must be followed, the standard works read, favorite problems worked out, short-cuts conned and rapidity of output exercised. Stokes, of Pembroke, had been senior wrangler in 1841; Cayley, of Trinity, in 1842; and Adams, of John's, in 1843; why not Thomson, of

Peterhouse, in 1845, argued Hopkins, who had the distinction of being second wrangler of the previous year?

But when the ordeal was over and the work of all candidates appraised, Thomson's name was second on the list, with Parkinson, of John's, at the top. Hopkins was disappointed, as he had a right to be, for it was thought by many and said by some that Parkinson was not fit to sharpen Thomson's pencils. At the examination for the Smith's prizes, which immediately followed, and which was generally regarded as a higher honor and a better test of original ability, the order was reversed, and Thomson's star blazed out with the brilliancy of the first magnitude.

We have here an instructive instance of the failure of an examination to place rightly the most gifted man; that of Sylvester, in 1837, and Clerk Maxwell, in 1854, both of whom were second wranglers, are equally so. Examinations, however, seldom fail in justly rating candidates when originality is not a necessary qualification, but only a sound knowledge and liberal interpretation of the subjects laid down in the syllabus; a good memory and rapidity of writing will do the rest.

Thomson committed the fatal mistake in the tripos examination of devoting too much time to a particular question in which he was deeply interested. It was a curious coincidence that the solution which Parkinson sent in to the same question was almost identical with that of his rival for mathematical honors. On being questioned about the matter by the Moderators, Parkinson said that he had read the solution some time before in the *Cambridge Mathematical Journal*; Thomson's explanation was that the solution given in the Journal was his! As he had not memorized the details, he was obliged of course to work the problem out *de novo*.

Parkinson in later years wrote a treatise on elementary mechanics that has long since made way for others; Thomson, on the other hand, published in collaboration with Tait a *Treatise on Natural Philosophy* for advanced students, which became at once the accepted standard. Throughout this treatise, the view is emphasized that physics deals with realities more than with theories, with mutual relations more than with their mathematical expression. Helmholtz thought so highly of this work that he translated it into German, saying in his preface: "William Thomson, one of the most penetrating and ingenious thinkers, deserves the thanks of the scientific world, in that he takes us into the workshop of his thoughts and unravels the guiding threads which have helped him to master and set in order the most resisting and confused material." And again: "Following the example given by Faraday, he avoids as far as possible hypotheses about unknown subjects, and endeavors to express by his mathematical treatment of problems simply the law of observable phenomena."

We are not to think of Thomson, the undergraduate, as of one who gave himself up, mind and body, to his favorite studies; he knew how to combine, in some measure, the *dulce* with the *utile*, for he was fond of music, and so proficient in the art that he was elected President of the Musical Society. He also took a practical interest in aquatic sports, and on the Cam he could ply his sculls with the best of the men. Indeed, he was fond of the water all through life, his *Lalla Rookh* being well known on the Clyde and in the Solent. Expert in the navigation of his yacht, he liked to be out on the deep, caressed by wind and buffeted by wave, on which occasions he usually studied, pencil in hand, problems connected with navigation and hydrodynamics.

Thomson was never without his note-book. Even in his journeys to London, when he usually took the night train to save time, his mind was active, and the green-book was in frequent requisition to receive thoughts that occurred relative to problems that engaged his attention. Unlike many mortals, he was able to sleep soundly on those night trips, although in the early days he had none of the luxuries of traveling which we consider indispensable to our comfort.

Helmholtz records that, being on the *Lalla Rookh* on one occasion, Thomson "carried the freedom of intercourse so far that he always had a mathematical note-book with him; and as soon as an idea occurred to him, he began to reckon right in the midst of company." This reminds us of the answer which Newton gave to a friend who asked him how he accomplished so much. "By constantly thinking of it," was the brief reply. Concentration of the faculties is necessary for all good work; a distracted mind never achieved anything of value in philosophy, in science, in religious worship. Concentration is like a convex lens, which brings rays to a focus; whereas distraction is like a concave lens, which breaks them up into a number of divergent and scattered elements.

On leaving Cambridge in 1845, Thomson proceeded to London, and was warmly received by Faraday, then of world-wide reputation. He next went to Paris, where, in the laboratory of Regnault, he devoted himself to original research, under the direction of that great and accurate physicist who was then carrying out his classic work on the thermal constants of bodies.

The year 1846 marks an epoch in Thomson's life; for, in that year, he was chosen to succeed Nichol, his friend and master, in the chair of natural philosophy in the University of Glasgow. Though only in his twenty-second year, he chose for the subject of his inaugural address the age of the earth, a subject which continued to have a life-long interest for him because of its very fascination, and perhaps, too, because of the opposition which his views aroused on the part of biologists and geologists. These demanded untold æons for the original fire-mist to cool down and form a spinning globe fit to be the abode of organic life, whereas Thomson endeavored to show the weakness of the arguments which they advanced to uphold their claim for unlimited time. Basing his estimate on the rate of increase of temperature as we go below the earth's surface, he concluded that the earth required from 100 to 200 million years, and probably less, to cool from its molten state to its present condition.

Impressed by the value of the experimental work which he did under Regnault in Paris, Prof. Thomson gave himself no rest until he secured a place in which the demonstrations of the lecture-room could be supplemented by qualitative and quantitative work in the laboratory. This was the first "physical laboratory" open to students in Great Britain, a fact that makes the year 1846 a memorable one in the history of university development. Two apartments were allotted him for experimental purposes, *viz.*, an abandoned wine-cellar and a disused examination-room, to which, as time went on, were added a corridor, some spare attics, and even the university tower itself, so great was the power of annexation possessed by the young Professor. In those dark and cheerless rooms, a few old instruments were installed, after which students were invited and work begun. A band of men, whose ardor was enkindled by the glowing enthusiasm of the presiding genius, gathered around him, and helped him to carry out investigations on the properties of metals, on moduli of elasticity, elastic fatigue and atmospheric electricity. Among this band of earnest students it will suffice to mention the names of the late Prof. Ayrton, an eminent electrician; Prof. John Perry, known for his Homeric battles in favor of reform in the teaching of mathematics; Sir William Ramsay, the discoverer of the "newer" gases of the atmosphere; and Prof. Andrew Gray, who succeeded his master in the University of Glasgow.

Writing of his laboratory experiences, Prof. Ramsay says: "I remember that my first exercise, which occupied over a week, was to take the kinks out of a bundle of copper wire. Having achieved this with some success, I was placed opposite a quadrant electrometer and made to study its construction and use." "Although this method," he adds, "is not without its disadvantages—for systematic instruction is of much value—there is something to be said for it. On the one hand, too long a course of experimenting on old and well known lines is likely to imbue the young student with the idea that all physics consists in learning the use of apparatus and repeating measurements which have already been made. On the other hand, too early attempts to investigate the unknown are likely to prove fruitless for want of manipulative skill and for want of knowledge of what has already been done."

Prof. Gray wrote: "In the physical laboratory, Prof. Thomson was both inspiring and distracting. He continually thought of new things to be tried, and interrupted the course of work with interpolated experiments which often robbed the previous sequence of operations of their final result."

It may bring a grain of consolation to teachers who meet with troublesome elements in the discharge of their duties, to know that Thomson, great and brilliant as he was, had similar experiences now and again. At one time a book of mathematical data would be removed from the place assigned to it, upon which he would give orders that it should be chained to the table; at others, there would be no chalk near the blackboard, and then the assistant would be solemnly instructed to have one hundred pieces available next time. On one occasion, he settled in a very novel manner the case of a student who insisted on disturbing the class by moving his foot back and forth on the floor. Calling his assistant, Thomson told him in a whisper to go down into the room under the tiers of seats, to listen attentively, and locate the wandering foot by its distance from two adjacent walls of the building. On his return to the lecture-room, the triumphant assistant gave the desired coordinates to the Professor, who took out his tape at once and measured off the distances, by which the outwitted offender was mathematically located. In obedience to orders, the latter rose and left the room, muttering a few graceful epithets as he went, in honor of Descartes, the founder of a system of geometry that could serve so well the twofold purpose of the detective and the mathematician.

It was the custom in Glasgow to open the daily sessions, morning and afternoon, with prayer, the selection of which was left to the discretion of the Professor. Thomson usually recited from memory the third collect from the morning service of the Church of England, to which he sometimes added reflections of his own for the spiritual benefit of his hearers.

In his teaching, Prof. Thomson was particularly insistent that his students should not bow their intellects in mute admiration before an array of mathematical symbols; but that, on all occasions, they should seek the physical meaning behind them. Writing on his blackboard one day *dx/dt*, he was not satisfied when told that it represented the ratio of the increment of x to the increment of the independent variable t (time); he wanted the student to say it represents velocity. He himself was so wont to look for the physical meaning of symbols that, like the prophets of old, he saw many things that were hidden from the eyes of ordinary mortals.

He had the rare gift of translating mathematical equations into real facts; and he strove all throughout his life, by word and writing, to purify mathematical theory from mere assumptions. He often said that he could not understand a thing until he was able to make, or at least conceive, a model of it.

He had a "keen mathematical instinct," as Prof. Silvanus P. Thomson puts it in a letter to the writer, an insight that "grew to see things." He often left matters in the dark for years, then returned to see them in the clear light of truth. At the age of sixteen, he wrote a mathematical essay on the figure of the earth; and at eighty-three, took it up again in order to add a note to the argument!

Thomson was discursive in his lectures, and was never able to boil the matter down to suit the taste and digestive powers of the ordinary student. The activity of his mind and its fecundity were such that new ideas, new problems, new modes of treatment were continually occurring, and with such fascination that he would leave the main subject to indulge in what often proved prolonged digressions. One of his bugbears was our system of weights and measures, which he denounced in season and out of season as "insane," "brain-wasting" and "dangerous." Occasionally epithets of a more caloric nature would escape the lips of the indignant Professor, who, as a consequence of his denunciation, had always to be indulgent to students who chanced to be shaky in the matter of Troy weight, avoirdupois weight or even apothecaries weight.

In later years, I heard Lord Kelvin at the Royal Institution, London, on some of his favorite dynamical subjects, such as the gyrostat, vortex rings and the like. However impressed by his keen eye, intellectual forehead, his mastery of the subject and wealth of illustration, I was no less impressed by his vivacity, his enthusiasm and the rapidity with which he could leave a train of thought and return to it again.

At meetings of the British Association, he always had something illuminating to say; but not infrequently, carried away by a torrent of ideas, he would indulge in a superfluity of detail, forgetting that other speakers had to be heard and other papers read.

The idea of connecting the Old World with the New by means of an electric cable laid on the bed of the ocean, seemed to most people in the 'fifties quixotic and utopian. Manufacturers said such a cable could not be made; engineers, that it could not be laid; electricians, that it could not be worked; and financiers, that if laid and worked, it would never pay. But with a Field to look after the financial interests of the scheme, and a Thomson to attend to electrical quantities, there was no tilting at windmills, and the utopian scheme became in due time the cable whose core pulsated with the news of the world.

As early as 1850, Bishop Mullock, of St. John's, N. F., addressed to an American newspaper, called the *Courier*, a letter in which he advocated a telegraph line from Newfoundland to New York, so that the news of mail steamers could be intercepted and wired to that City. In 1852, the "Newfoundland Electric Telegraph Company" was formed for the purpose of carrying out a similar plan. This was to be accomplished by means of a telegraph line from Cape Race, at the eastern extremity of Newfoundland to Cape Ray, on the western, as well as by short cables over to Cape Breton Island, to Prince Edward Island and the mainland, and thence by ordinary telegraph lines to Canada and the United States. But owing to the want of money, nothing was done.

The first attempt at laying a cable under the Atlantic was made by the Atlantic Telegraph Company in 1857, after a careful survey of the ocean had revealed the existence of a submarine plain, or extended table-land, on which the cable could rest undisturbed by passing keels, monsters of the deep or angry billows. The result was the first of a series of failures, which caused great perplexity and depression at the time; for, after 330 miles had been paid out from Valentia on the Irish coast, the cable suddenly parted, burying in 2000 fathoms of water an electrical conductor which had cost $150,000 for its manufacture.

A second attempt was made in 1858, when the U. S. frigate *Niagara* and H. M. S. *Agamemnon,* each carrying half of the cable, met in mid-ocean; and, after splicing the two ends together, steamed away in opposite directions, the *Niagara* toward Newfoundland and the *Agamemnon* toward Valentia. Fortunately for the enterprise, Prof. Thomson was on board the English ship as chief electrician. No doubt, his mind turned many a time during those anxious days to Fourier's differential equation for the flow of heat along a conductor, and his own application of it to the conduction of the electric current through the copper core of the cable as it came up from the tanks, trailed out behind the ship, dipped silently into the blue water and slowly settled down to its bed of ooze on the ocean floor.

After a series of disheartening mishaps, necessitating as many returns of the ships to the rendezvous in mid-ocean, the *Agamemnon* landed the shore-end safely in Valentia; and the *Niagara,* after rolling and pitching for days and nights in tempestuous seas, landed hers in Trinity Bay on the morning of August 5th, 1858, on which historic date the telegraphic union of the two worlds was finally consummated and the great feat of the century accomplished.

Though not fully realized at the time by the capitalists who financed his scheme, by the engineers and electricians who carried it out, or even by statesmen, economists and social reformers, the slender copper cord, buried away from human ken amidst the *débris* of minute organisms, was destined to effect a revolution in the affairs of men greater than any achieved by the wisdom of sages or the policy of legislators.

Owing to the electrostatic capacity of the cable, signaling would have been difficult and unsatisfactory had it not been for the resourcefulness of Prof. Thomson, who devised his reflecting galvanometer to serve as receiving instrument. The principle of the mirror applied in this way was not new, for it had been suggested by Poggendorff and even used by Gauss in connection with very heavy magnets. The magnets used by Thomson, on the other hand, were strips of watch-spring weighing about a grain each, so that even a very weak current coming through the cable would be sufficient to produce strong displacements of the spot of light on the scale. Thomson was clearly the first to insist on small dimensions in magnetic instruments, and to show that reduction in size would be attended with corresponding increase in sensitiveness.

The mirror galvanometer, surrounded with a thick iron case to screen it from the magnetic field due to the iron of the ship, the "iron-clad galvanometer" as it was called, was used for the first time on the telegraphic expedition of 1858.

The instrument itself, which was fitted up on board the *Niagara* and which was connected with so many episodes of thrilling interest, was placed by Prof. Thomson in the collection of historical apparatus in the University of Glasgow, where it is at the present day.

Beautiful as was the invention of the mirror galvanometer, it gave neither warning of the beginning of a message nor a permanent record of it. Sitting in his dark room, the operator had to be always on the alert for the first swing of the spot of light over the scale. To obviate these drawbacks, Thomson, after some thinking and more talking with his friend White, of Glasgow, finally patented the *siphon-recorder*, in which a glass siphon of capillary dimensions is pulled to the right or left by the action of the current flowing through a light movable coil, and is thus made to register signals in ink on a vertical strip of paper which is kept in uniform motion by a train of clockwork. It is by this simple but very ingenious instrument that messages are received and recorded to-day at all the cable-stations of the world.

The inaugural message through the cable came from the Directors of the Atlantic Telegraph Company in Great Britain to the Directors in America, saying: "Europe and America are united by telegraph; glory to God in the highest, on earth peace and good will toward men."

The message from Queen Victoria to President Buchanan, consisting of 95 words, took 67 minutes in transmission; it read:

"The Queen desires to congratulate the President upon the successful completion of this great international work, in which the Queen has taken the deepest interest.

"The Queen is convinced that the President will join with her in fervently hoping that the electric cable which now connects Great Britain with the United States will prove an additional link between the nations whose friendship is founded upon their common interests and reciprocal esteem.

"The Queen has much pleasure in thus communicating with the President, and renewing to him her wishes for the prosperity of the United States."

The reply of President Buchanan was as follows:

"The President cordially reciprocates the congratulations of Her Majesty, the Queen, on the success of the great international enterprise accomplished by the science, skill and indomitable energy of the two countries. It is a triumph more glorious, because far more useful to mankind, than was ever won by conqueror on the field of battle.

"May the Atlantic telegraph, under the blessing of Heaven, prove to be a bond of perpetual peace and friendship between the kindred nations, and an instrument destined by Divine Providence to diffuse religion, civilization, liberty and law throughout the world. In this view will not all nations of Christendom spontaneously unite in the declaration that it shall be forever neutral, and that its communications shall be held sacred in passing to their places of destination, even in the midst of hostilities?"

The historian of the enterprise was Mr. John Mullaly, of New York, who was on the *Niagara* as secretary to Prof. Morse and subsequently to Mr. Cyrus W. Field and correspondent of the *New York Herald*. He has published three interesting works on the subject: a *Trip to Newfoundland, with an account of the laying of the submarine Cable* (between Port au Basque and North Sydney), 1855; *The Ocean Telegraph*, 1858; and *The first Atlantic Telegraph Cable*, a pamphlet of 28 pages, reprinted from the "Journal of the Franklin Institute," 1907. From it, we learn that Archbishop Hughes was one of the principal American subscribers to the capital of the Atlantic Cable Company.

When, in 1855, the subject of laying a cable under the Atlantic ocean began to be seriously considered, Thomson, who was then only 31 years of age, discussed in a series of masterly papers the theory of signaling through such conductors, showing *inter alia* that the instruments used on land-lines would be inoperative on cables, and also that the same speed of transmission could not be attained on cables as on ordinary telegraph lines. It was shown at the same time, that these differences are due to the fact that, unlike an air-line, the cable is an electrical *condenser* in which the copper core is separated from the waters of the ocean by a layer of gutta percha, a nonconducting material. As a submerged cable is, therefore, a long Leyden jar of great electrical capacity, it follows that a signal sent in at the American end will not reach the other instantly; for while the current flows along the conductor, it has also to charge up the cable as it progresses, which operation retards the signals, and also deprives them of the clearness and sharpness with which they were sent. The phenomenon is analogous to the diffusion of heat along a bar, the temperature of the various cross-sections rising in gradual succession until the distant end is reached. The mathematical investigations of Thomson showed the necessity of working slowly, and of using weak currents as well as very delicate receiving instruments. The interval of time required for the transmission of a signal from Newfoundland to Valentia is about one second.

Some years later, in 1858, Thomson had the opportunity of putting his theoretical views to the test of experiment on a grand, commercial scale, and had the satisfaction of finding that all his conclusions were confirmed. Electricians of the early period distrusted the inexperienced young man who had never erected a mile of telegraph line or even served for a month in a telegraph office; but their distrust was followed by admiration when they saw the efficient manner in which he handled every problem and dealt with every difficulty that occurred while laying the cable of 1858. It was generally admitted that, had it not been for the brilliant work of the young Glasgow Professor, many years would have passed away before the Old World and the New would have been brought into telegraphic communication.

Like all interested in the enterprise, Thomson was greatly shocked when the news reached him that signals could no longer be transmitted through the cable, which, after costing so much money, so much thought and labor, now lay a useless thing in two and a half miles of water. Attempts were made to raise it, but without success.

During its short life of less than a month, 366 messages were flashed through the cable, aggregating 4359 words of 21,421 letters.

The failure of the pioneer cable has been attributed to a variety of causes, chief of which were defective construction and imperfect paying-out machinery, which produced unequal strains in the cable. Defective as the cable was at the moment of immersion, the various troubles became intensified with time, until at last, when provoked by the feebleness of the signals, the injudicious electrician at Valentia had recourse to the great penetrative power of the induction coil, and gave the dying cable the *coup de grâce*.

An experiment made by Mr. Latimer Clark is not only germane to the subject, but is also of very great interest. Writing from Valentia on Sept. 12th, 1866, Mr. Latimer Clark says: "With a single galvanic cell, composed of a few drops of acid in a *silver thimble* and a fragment of zinc, weighing a grain or two, conversation may easily, though slowly, be carried on through one of the cables (1865, 1866) or through the two joined together at Newfoundland; and although in the latter case, the spark, twice traversing the breadth of the Atlantic, has to pass through 3700 miles of cable, its effects at the receiving end are visible in the galvanometer in a little more than a second after contact is made with the battery. The deflections are not of a dubious character, but full and long, the spot of light traversing freely a space of 12 in. or 13 in. on the scale; and it is manifest that a battery many times smaller would suffice to produce similar effects."

Not to be outdone by the English electrician, Mr. William Dickerson devised the gun-cap cell, which he used in 1866 with success in transmitting signals from Heart's Content, Newfoundland, to Valentia on the Irish coast.

A piece of No. 16 bare copper wire was procured, one end of which was firmly twisted around the head of an empty *percussion-cap*. To one end of another similar length of wire was bound, with fine copper wire, a short strip of zinc bent at a right angle to form the anode element of the diminutive cell. After charging the cell with a drop of acidulated water of the size of an ordinary well-formed tear, and properly connecting the terminals with earth and cable, signals were transmitted over the cable by the infinitesimal current generated by this novel cell. The receiving operator reported that the signals were "awfully small"; but they were intelligible, and messages were successfully transmitted under the ocean by this tiny element.

Contrast with this Lilliputian cell the enormous power that was used on the cable of 1858 toward the end of its short existence, when batteries of 380 and 420 Daniell cells were employed to force signals across.

When, in 1865, it was decided to make another attempt at laying a cable under the Atlantic, Prof. Thomson, whose reputation was enhanced during the seven intervening years by a number of communications on the theory and practice of submarine telegraphy, was again retained as scientific expert in a consultative sense, with Mr. Cromwell F. Varley as chief electrician. In accordance with the costly experience that had been gained, a new cable was made and coiled on board the *Great Eastern*, a leviathan which was well fitted for the work by the great manœuvring power afforded by its screw and paddles combined. Leaving Valentia, the big ship steamed with her prow to the west at a slow rate of speed, in order to give the cable time to sink beneath the waves and adapt itself to the configuration of the ocean floor. Eleven hundred miles had been successfully paid out when, to the consternation of all, the cable suddenly snapped and disappeared in more than two miles of water. Attempts were made during the next nine days to recover it from those abysmal depths; and, though grappled many times during those trying hours, it gave way each time under the strain to which it was subjected. Like its predecessors of 1857 and 1858, the cable of 1865 was finally abandoned to its fate, and the *Great Eastern* returned home with three greatly disappointed men on board, *viz.*, Prof. Thomson, Mr. C. F. Varley and Captain (later Sir James) Anderson.

In the following year, a sum of three-quarters of a million sterling, nearly $4,000,000, was offered to the Directors of the "Telegraph Construction Company" if they would complete the cable of 1865 and lay a new one. After consultation and careful consideration, the offer was accepted and the cable constructed according to the best engineering knowledge available.

In 1866, Prof. Thomson was again on board the *Great Eastern* with Captain Anderson; and this time the big ship had snugly coiled up in her deep, cavernous tanks *the* cable that was destined to put Europe and America in permanent telegraphic communication. With a well-manufactured cable, improved paying-out machinery and an experienced staff of mechanical engineers, not to mention the foremost electricians of the day, the immersion of the cable was successfully effected, after which the American end of the cable of 1865 was raised, a new length spliced on, and the shore-end safely landed in Trinity Bay. Europe and America were thus united together by two electric bonds.

It may here be mentioned that ocean cables are usually made in three sections, called, respectively, the shore-end, the intermediate section and the deep-sea section. It is clear that the submerged conductor needs the greatest protection in the shallow water that surrounds the coast, where it lies on a pebbly or rocky bottom, exposed to the drifting action of currents and tides, as well as to the haling flukes of the anchors of storm-tossed ships. In deep water, on the other hand, there is neither shingly bottom nor violent movement to displace and abrade the cable; for all is quiet and peaceful in the profound depths where the god of the trident holds his court; and hence few coverings and a light armor afford sufficient protection. The wear and tear in the ocean depths is a vanishing quantity when compared with the abrasive effects near coast-lines. Looking at the sections of an ocean cable, the biggest and heaviest is the shore-end, while the thinnest and lightest is that which goes down into the depths of the sea. The lengths of the various sections are determined by the survey of the route, which is always carefully made before completing the specification of the cable. Moreover, as the position of the cable-ship at noon every day is known from its longitude and latitude, it follows that the location of the cable on the bed of the ocean is also exactly known. When a cable is broken either by an upheaval or by a subsidence of the ocean floor, the distance of the rupture from the shore end is determined by an electrical test, after which a repair-ship is dispatched to the spot, when the cable is lifted, the "fault" cut away, a new length spliced on, and the amended cable allowed to settle down into its watery depths.

At the present time (July, 1909), there are sixteen cables carrying the work of the North Atlantic, at an average speed of 20 words a minute duplex, or 40 words a minute, counting both directions.

This cable narrative affords as striking an illustration of the *triumph of failure* as any recorded in the history of human enterprise. It was a victory of mind over matter; of character and tactfulness, energy and endurance over difficulties of every kind, moral and financial, mechanical and meteorological. The four expeditions of 1857, 1858, 1865 and 1866 represent years of hard work, anxiety and distressing failures; but, sustained by the patience of hope and by an unshaken confidence in the soundness of the enterprise as well as in the ability of their staff, the Directors of the Atlantic Company were well rewarded for the disappointment occasioned and the monetary losses incurred. "It has been a long struggle," said the initial promoter of the enterprise, Mr. Cyrus W. Field, speaking at a banquet given in his honor on November 15th, 1866, at the Metropolitan Hotel, New York, "a long struggle of nearly thirteen years of anxious watching and ceaseless toil. Often my heart was ready to sink. Many times, when wandering in the forests of Newfoundland in pelting rain, or on the decks of ships in dark, stormy nights, I almost accused myself of madness and folly to sacrifice the peace of my family for what might have proved but a dream. I have seen my companions, one after another, fall by my side, and I feared that I, too, might not live to see the end. And yet one hope has led me on; I prayed that I might not taste of death till the work was accomplished. That prayer has been answered; and now, beyond all acknowledgments to men, is the feeling of gratitude to Almighty God."

It was men like Field and Thomson that the poet had in mind when he wrote:

The wise and active conquer difficulties By daring to attempt them. Sloth and folly Shiver and shrink at sight of toil and labor, And make the impossibility they fear.

Shortly after his return home, Prof. Thomson was knighted for his splendid services in connection with sub-oceanic cables, and was also honored with the freedom of the City of Glasgow.

If while journeying over land or sea, Sir William's mind was always active, his eyes were also open and observant. In the numerous voyages which he undertook in the interest of cable companies, he seems to have been struck by the unreliable character of the ordinary apparatus used in taking soundings, consisting of a heavy weight suspended by a thick hempen cord unwound from a reel. Owing to the massiveness of the cord, the motion of the ship and currents in the water would necessarily deflect it from the vertical, so that the soundings recorded would be in excess of the true depth. To remedy this defect, Thomson replaced the rope, at first by a steel wire, and later by a thin strand of steel wires, on which the speed of the ship has but little effect; the sinker descends vertically with considerable velocity, and is raised with equal rapidity by suitable winding-up machinery placed in the stern of the ship. The sinker carries a gauge consisting of a quill-tube open at the lower end and closed at the top.

The inside, which is coated with silver chromate, shows by the discoloration produced by the action of the sea water how far the water has compressed the air in the tube. By comparison with a graduated ruler, the depth is then read off. When the sinker reaches bottom, the heavy weight is detached automatically, so that there is but little strain on the wire as it ascends with its thermometer and battery of tubes containing samples of the depths reached.

A story is told in connection with this sounding-machine which shows the vivacity and wit of the inventor. Having brought his friend Joule into White's one day, he pointed to a number of coils of steel wire lying on the floor, informing his English friend of "mechanical-equivalent" fame at the same time that he intended the wire for sounding purposes. Upon Joule's innocently asking what note it would sound, he received the prompt answer, "the deep sea"!

Another subject to which Sir William gave some attention after his experiences on the ocean is the navigating compass. His observations led him to distrust the long, heavy needles then in general use on shipboard. Besides the friction to which the pressure on the pivot gives rise and which necessarily diminishes the sensitiveness of the needle, there was another objection, due to the difficulty experienced in successfully applying steel magnets and soft-iron masses to compensate for the magnetism of the ship and for the changes induced in it by change of place in the earth's magnetic field.

As a result, Prof. Thomson devised a compass-card which is remarkable for its lightness and sensitiveness. It is made of two sets of magnets, containing four needles each, arranged symmetrically on the right and left of the pivot. The four needles, forming a set, are of unequal length, ranging from 3-1/4 to 2 inches, with the shortest outermost. Such a card, with its associated correctors of steel magnets and soft-iron balls, has added greatly to the safety and certainty of navigation; and as such, it is used to-day in the merchant service and in the navies of most countries of the world.

As we have seen, Thomson had the keen, racy wit of his race. Lecturing before the members of the Birmingham and Midland

Institute in 1883, he placed himself and his nationality on record in a very humorous way. His subject was "The six gateways of Knowledge." As will be remembered by the readers of *The Pilgrim's Progress*, old Bunyan likened the soul to a citadel on a hill having no means of communication with the outer world save by live gates, *viz.*, the eye gate, the ear gate, the mouth gate, the nose gate and the feel gate. These are the five senses by which we obtain our knowledge of the material world which surrounds us. But Prof. Thomson took issue with Bunyan, with Reid, and the metaphysicians of all time in maintaining in this lecture that we have six gateways of knowledge instead of five, justifying the position which he took by affirming that the sense of touch is really twofold, one of heat and the other of force. It does not appear, however, that he made any marked impression on the philosophic thought of the day, for psychologists continued to write with undisturbed equanimity of the five senses and not the six.

It was on this occasion that Prof. Thomson said: "The only census of the senses, so far as I am aware, that ever before made them more than five was the Irishman's reckoning of seven senses. I presume the Irishman's seventh sense was common sense; and I believe that the possession of that virtue by my countrymen, *I speak as an Irishman,* I say the large possession of the seventh sense which I believe Irishmen have, will do more to alleviate the woes of Ireland than the removal of 'the melancholy ocean' which surrounds its shores."

For the successful operation of cables, telegraph lines and scientific investigations of all sorts, a system of practical electrical units, accepted by all companies and countries of the world, was soon found to be indispensable. The pioneer in the movement for establishing an international system of electrical standards was Mr. J. Latimer Clark, who, assisted by his distinguished partner, (Sir) Charles Bright, prepared a paper on "The formation of Standards of Electrical Quantity and Resistance," which was read at the Manchester meeting of the British Association in 1861. Prof. Thomson was present; and, at his instance, a committee was appointed to report on the general question of electrical units. This was the first meeting of a committee that was destined to accomplish much in the electric and electromagnetic field; it was the initial impulse of a movement that brought renown to the entire body of English electricians. Such units as the ohm, the volt and the farad met with immediate acceptance, while later on the ampere, the coulomb, the watt and the joule were introduced. Among the members of this body besides Prof. Thomson, were such able men as Clerk Maxwell, Joule, Lord Rayleigh, Sir William Siemens, Johnstone Stoney, Balfour Stewart, and Carey Foster.

The world is then indebted to the insistence and advocacy of Prof. Thomson for the general acceptance of the "C.G.S." system of measurement, which involves the centimeter (length), the gram (mass), and the second (time) as the fundamental units from which all others are derived.

Prof. Thomson has claims in the "wireless" field also; for as far back as 1855, he studied the nature of the discharge of a condenser and proved mathematically that, under certain conditions easily realized in practice, such discharges are of an oscillatory character, consisting of a forward and a backward rush of electricity between the two coatings of the condenser. As pointed out on page 92, Prof. Henry had reached the same conclusion in 1842, and Helmholtz in 1847; but Thomson's insight into the phenomenon is keen and his mathematical analysis of it very remarkable.

Just as the to-and-fro motions of the prongs of a tuning-fork give rise to sound-waves in the air, so the electric oscillation due to a condenser discharge sets up in the universal ether electric waves which flash the news of the world over continents and oceans with unthinkable velocity.

By special request, Sir William Thomson gave, in 1884, a course of lectures at the Johns Hopkins University, Baltimore, to an audience of "professional fellow-students in physical science," as he called the *élite* of American men of science, twenty-one in number, assembled to hear him. These accomplished physicists he also affectionately called his "twenty-one coefficients."

The subject was the wave-theory of light, and the object of the lecturer was to show how far the phenomena of light, such as its transmission, refraction and dispersion, could be explained within the limits of the elastic solid theory of the ether, which makes that hypothetical medium rigid, highly elastic and non-gravitational. From the very first lecture, Sir William assumed a cold and diffident attitude toward the rival theory of Clerk Maxwell, which makes light an electromagnetic phenomenon; and though his own presented formidable difficulties, and its rival was universally accepted, the veteran Professor assured his hearers that the elastic solid theory is the "only tenable foundation for the wave-theory of light in the present (1884) state of our knowledge."

Despite the energy which he displayed, his luminous argumentation and close logic, Kelvin made no converts among his "twenty-one coefficients"; and it soon became evident that he was championing a lost cause. Newton did the same when he held tenaciously to the corpuscular theory of light; and in doing so, let it be said, that he retarded the acceptance of the wave-theory and the advance of science by a hundred years.

A few years after the Baltimore lectures, official recognition of his distinguished services and of his eminence in science came to Sir William Thomson when, in 1892, he was raised to the peerage, with the title of Baron Kelvin of Netherhall, Kelvin being the name of a stream which passes near the buildings of the University of Glasgow and flows into the Clyde, while Netherhall is that of his country-seat at Largs, in Ayrshire, 40 miles from Glasgow.

As to the structure of matter, Kelvin lived to see the "atom" of his youth and mature years shattered into fragments, and the atomic theory of matter rapidly yielding to the electronic. Though he maintained an open mind toward the new school of physics, he was reserved and conservative toward the revolutionary doctrine of extreme radio-activists. He did not believe in the transformation of one elementary form of matter into another; and he strenuously combated the theory of the spontaneous disintegration of the atom.

Notwithstanding a long life devoted to the study of mathematical and experimental physics, during which Kelvin unraveled many a difficult problem in electricity and magnetism and added many a beautiful skein to the texture of our knowledge in electrostatics and electrokinetics, that illustrious man, the acknowledged leader in physical science, made a public admission in 1896 which caused a great stir throughout the scientific world. It was on the occasion of the celebration of the golden jubilee of his professorship of natural philosophy in the University of Glasgow. Delegates had come from all parts of the world; kings and princes had sent their representatives; universities and learned societies of every country of the Old World and the New vied with one another in doing honor to the scientist who had figured so long and so conspicuously in the advances of the age. It was on that solemn occasion and in presence of such a notable assembly that Kelvin made the astonishing admission that, although he had been a diligent student of electricity and magnetism for a period exceeding fifty years, and although he had pondered every day for forty years over the nature of the ether and the constitution of matter, he knew no more about their essence, about what they really are, than he knew at the beginning of his professional work.

This confession, remarkable by reason of the man who made it and the circumstances in which it was made, has always appeared to the writer of these lines as having more of the ring of disappointment in it than of blank failure. Kelvin's great analytical mind early and persistently strove to penetrate the closely guarded secrets of nature; and because Dame Nature did not yield to his open sesame, but persisted in her reticence, the philosopher grew pessimistic and disappointed; and, under the sway of such feelings, he summed up the result of his life-quest after the ultimate problems in science and pronounced it a "failure."

A "failure" it was not, if science is the discovery and registration of the laws of God as revealed in the universe of mind and matter; for few men of his generation, if any, made more contributions of the first order to the theory of electrostatics, to the doctrine of energy, to hydrodynamics and the thermo-electric properties of matter. This note of disappointment, or wail of despondency, had been sounded before by Faraday, who said that, the more he studied electrical phenomena, the less he seemed to know about electricity itself. Was not Laplace animated by a kindred feeling when he spoke about the infinitude of our ignorance? Lastly, was not this intense feeling of our limited powers precisely that which, after all his discoveries in mathematics, in optics and in celestial mechanics, made Newton compare himself to a child standing on the beach with the vast ocean of truth before him, unfathomed and unexplored?

Kelvin gave a beautiful example to the world when, after resigning the chair which he had occupied for fifty-five years in the University of Glasgow, he immediately proceeded to enter his name on the undergraduate list, intimating by such an act that, whether a man is a professor-in-ordinary of natural philosophy or a professor emeritus, he must ever be a student, in close touch with nature.

Lord Kelvin had the happiness of enjoying good health throughout all the years of his long career, a happiness due in part to nature, and in part also to the simplicity, frugality and regularity of his life.

As already said, he was fond of cruising in European waters in his yacht *Lalla Rookh* during the summer months, and even venturing out on the Atlantic as far as Madeira, for,

He loved the sea, and what is more, He loved it best when far from shore.

In later years, however, owing to facial neuralgia, he was accustomed to spend a month or so every summer with Lady Kelvin at Aix-les-Bains, from which visits he always derived much benefit.

While making some experiments in a corridor of his beautiful home at Netherhall, he caught a chill on November 23d, 1907, from which he never rallied, despite the cares and attentions that were fondly lavished upon him. The bulletins that were issued concerning his condition were read all the world over with more concern than if they referred to a reigning sovereign or an heir apparent. Every teacher of physics, mathematical or experimental; every man interested in the advance of science and the spread of knowledge, anxiously awaited news from the sick-room of the illustrious patient—news that was transmitted to the ends of the earth by the siphon-recorder invented by the dying scientist in the heyday of his life; and when the word came that Kelvin had breathed his last, that cablegram brought universal sorrow for the quenching of the brightest light of the age and the loss of the leading scientist, the model man and faithful Christian.

It was in the fitness of things that the man who was considered the greatest since Newton should be buried in Westminster Abbey, and that the mortal remains of Lord Kelvin should find a resting-place next to the grave of the genius who thought out the *Principia* and discovered the gravitational law which governs the planetary as well as the stellar universe.

If asked to say what impressed me most in Lord Kelvin, I would mention the cordial manner in which he welcomed those who sought advice; the encouragement which he held out to students; his absolute devotion to truth; his fair-mindedness and candor; his reverence in dealing with the problems of the soul and the destiny of man; and the uniform, tranquil happiness of his life, due, under God, to his profound religious belief and noble Christian life.

A man of strong convictions, Kelvin did not, however, wear his religion on his sleeve, but treasured it in the depths of his heart, where it was never disturbed by the tossing and ever-changing wave-forms of individual opinion. He quietly but uniformly maintained that physical science demands the existence and action of creative power; and he did not shrink from affirming this conviction whenever circumstances seemed to require it, as was the case on the memorable occasion of his Presidential address to the members of the British Association in 1871. In concluding that brilliant discourse, he said: "But strong, overpowering proofs of intelligent and benevolent design lie all around us; and if ever perplexities, whether metaphysical or scientific, turn us away from them for a time, they come back upon us with irresistible force, showing to us, through nature, the influence of free will, and teaching us that all living beings depend on one ever-acting Creator and Ruler."

Once when particularly disgusted with the materialistic views of those who, while denying the existence of a Creator, attributed the wonders of nature, animate and inanimate, to the potency of a fortuitous concourse of atoms, he wrote to Liebig, asking him if a leaf or a flower could be formed or even made grow by chemical forces, to which he received the significant reply from the famous chemist of Giessen: "I would more readily believe that a book on chemistry or on botany could grow out of dead matter by chemical processes."

We have already referred to the custom which obtained in the University of Glasgow, of beginning the daily sessions by invoking the blessing of heaven on the work about to be undertaken. Having liberty in the matter of choice, Prof. Thomson selected for this purpose a prayer from the morning service of the Church of England, which reads: "O Lord, our heavenly Father, almighty and everlasting God, who hast safely brought us to the beginning of this day; defend us in the same with Thy mighty power; and grant that this day we fall into no sin, neither run into any kind of danger; but that all our doings may be ordered by Thy governance, to do always what is righteous in Thy sight; through Jesus Christ, our Lord, Amen."

Academical honors were showered upon Lord Kelvin by seats of learning, ancient and modern; he was a D. C. L. Oxford, LL. D. Cambridge, and a D. Sc. London; he was President of the Royal Society from 1890 to 1895; President of the British Association in 1871; Knight of the Prussian Order *Pour le Mérite*, and Foreign Associate of the *Institut de France*.

His published works include a "Treatise on Natural Philosophy," 2 vols., written in collaboration with Prof. Tait, of Edinburgh (the two authors were often referred to as T and T'); "Contributions to Electrostatics and Magnetism"; "Collected mathematical and physical Papers," 3 vols.; "Popular Lectures and Addresses," 3 vols.; and the

"Baltimore Lectures." These, as well as the instruments which he devised for navigation, for the finest work of the laboratory, as well as for the commercial measurement of current, potential, and energy, form a monument to Lord Kelvin that will be *aere perennius*.